Advance Praise

Varun Aggarwal offers a compelling, data-driven, and analytical march through India's science landscape. While we might not get science to the stature of cricket or Bollywood in the Indian imagination, at least we have to try to make it much more discussed. There is no substitute for a laser focus on science to foster social and economic change.

—Tarun Khanna
Jorge Paulo Lemann Professor, Harvard Business School
and Director, South Asia Institute, Harvard University,
Cambridge, Massachusetts

India has the potential to solve its own problems using science and technology. In doing so, it has a shot at solving the world's problems too. This first-of-its-kind book discusses how the nation can achieve this by taking advantage of its unique strengths. Varun weaves data-based insights with stories to demystify the world of research and India's position in it. He cites several interesting examples from his days at MIT and as an entrepreneur developing research-led products out of India. Science and technology are not some idealistic "good to haves"; they have increasingly become instruments for progress, and Varun's thoughtful passion shows the way.

—Sanjay Sarma
Professor of Mechanical Engineering and Vice President for
Open Learning, Massachusetts Institute of Technology (MIT)

This book is a must-read for people who want to learn more about science in India, as it offers a comprehensive account of India's science and technology ecosystem. Varun examines the national policy on science and innovation, the workings of India's top research institutions, the role of industry, and the mind-set of India's research personnel. Meticulously researched and based on a wealth of new data comparing countries across various research metrics, this book marks the start of an important and long-overdue dialogue in India about its position in world science and technology.

—Pradeep K. Khosla
Chancellor, University of California, San Diego

Just like you cannot imagine the Silicon Valley without Stanford, this book highlights the need for innovation backed by scientific research! To catapult India to the global stage of innovation, we need world-class research institutions, empowered researchers who tackle challenging original problems, and a vibrant ecosystem for entrepreneurs. This book is for any individual who dreams of an India that creates globally competitive start-ups and pioneers new technologies.

—Akshay Kothari
Head, LinkedIn India

It is a pity that India's scientific output isn't commensurate with the talent present in the country. This book explains why and warns about the opportunity cost of not acting. Varun presents a first lesson in the much-needed mentoring at various levels and for various actors to change the course. He nails it with his discussion around the importance of picking relevant problems, multidisciplinary research, building end-to-end solutions, tight partnership between industry and academia, and nurturing of appropriate ecosystems. A must-read for students, professors, researchers, policymakers, corporate leaders, media, and even parents who continue to exercise great career-related influence on the future protagonists. Simple yet subtle, critiquing yet constructive, Varun tells an engaging narrative that comes straight from his heart while also leading by example at many places.

—Sumit Gulwani
Research Manager, Microsoft, USA and Inventor of
Flash Fill in MS Excel

Can India become the next scientific superpower? With hard facts and plenty of personal anecdotes, seasoned entrepreneur Varun Aggarwal argues that it's time to unleash India's intellectual potential. Better yet, he shows us what needs to be done. [This book] is a roadmap to India's future.

—**Thomas Barlow**
Global Research Strategist and Author

India has established a significant export-driven IT industry and has a large and growing number of start-ups. However, it is clear to Varun, and to others, that the foundation of this enterprise—the science and research base of the country—is weak. The scale, quality, and impact of India's research and research training effort have not kept pace with its global peers. In a rigorous, evidence-based study that is frank and hard-hitting, but also ultimately hopeful, Varun identifies the issues and points to potential solutions. This book should be essential reading for those who care about India's science and research policy.

—**Arun Sharma**
Deputy Vice-Chancellor (Research and Commercialization),
Queensland University of Technology

Great research passes through various stages—from idea to peer review and publication to application. Different kinds of support are required at each stage and if the ecosystem is healthy, each stage simultaneously feeds on and supports the others. The strength of this book is precisely that it takes such a holistic approach, which is absolutely required to analyze the gaps that exist in India. Further, the book links the research all the way up to start-ups and entrepreneurship. Varun is one of the few people who can authoritatively talk about this journey since he started in the world of ideas and from there has created a very successful global company. Since he has also lived and worked in several countries, he is able to put things in a comparative global perspective. I found his comparisons of India, China, and the United States very revealing. I highly recommend this book.

—Shailendra Mehta
President, Director, and Distinguished Professor—Innovation & Entrepreneurship, MICA, Ahmedabad

Technology is changing how economies are organized and how wars are waged, affecting people and nations across the world. To be a leader internationally, a country must take the lead in science and technology. In this important book, Varun shows why, in comparison to India's impressive economic achievements, the country's technological achievements lag far behind. His insights and his thoughtful analysis of what needs to be done require to be widely understood and should be publicly discussed.

—Anirudh Krishna
Edgar T. Thompson Professor of Public Policy and Political Science, Duke University, North Carolina

LEADING SCIENCE AND TECHNOLOGY: INDIA NEXT?

LEADING
SCIENCE
AND
TECHNOLOGY:
INDIA
NEXT?

VARUN
AGGARWAL

Los Angeles | London | New Delhi
Singapore | Washington DC | Melbourne

First published in 2018 by

SAGE Publications India Pvt Ltd
B1/I-1 Mohan Cooperative Industrial Area
Mathura Road, New Delhi 110 044, India
www.sagepub.in

SAGE Publications Inc
2455 Teller Road
Thousand Oaks, California 91320, USA

SAGE Publications Ltd
1 Oliver's Yard, 55 City Road
London EC1Y 1SP, United Kingdom

SAGE Publications Asia-Pacific Pte Ltd
3 Church Street
#10-04 Samsung Hub
Singapore 049483

Published by Vivek Mehra for SAGE Publications India Pvt Ltd, typeset in 11/14 pt Garamond by Fidus Design Pvt. Ltd., Chandigarh and printed at Chaman Enterprises, New Delhi.

Library of Congress Cataloging-in-Publication Data
Name: Aggarwal, Varun, author.
Title: Leading science and technology: India next? / Varun Aggarwal.
Description: New Delhi, India: SAGE Publications India, 2018. | Includes bibliographical references and index. |
Identifiers: LCCN 2017044366 (print) | LCCN 2017058804 (ebook) | ISBN 9789352805105 (ePub) | ISBN 9789352805099 (Web PDF) | ISBN 9789352805082 (pbk. : alk. paper)
Subjects: LCSH: Science–Research–India. | Technology–Research–India. | Science–Study and teaching–India. | Technology–Study and teaching–India. | Science and state–India. | Technology and state–India.
Classification: LCC Q183.4.I4 (ebook) | LCC Q183.4.I4 A44 2018 (print) | DDC 507.2054–dc23
LC record available at https://lccn.loc.gov/2017044366

ISBN: 978-93-528-0508-2 (PB)

SAGE Team: Manisha Mathews, Sandhya Gola, Madhurima Thapa, and Ritu Chopra

To
Jagadish Chandra Bose, a courageous inventor and discoverer, and
Vivekananda, the inspiration for all my work

Thank you for choosing a SAGE product!
If you have any comment, observation or feedback,
I would like to personally hear from you.

Please write to me at **contactceo@sagepub.in**

Vivek Mehra, Managing Director and CEO, SAGE India.

Bulk Sales

SAGE India offers special discounts
for purchase of books in bulk.
We also make available special imprints
and excerpts from our books on demand.

For orders and enquiries, write to us at

Marketing Department
SAGE Publications India Pvt Ltd
B1/I-1, Mohan Cooperative Industrial Area
Mathura Road, Post Bag 7
New Delhi 110044, India

E-mail us at **marketing@sagepub.in**

Get to know more about SAGE

Be invited to SAGE events, get on our mailing list.
Write today to **marketing@sagepub.in**

This book is also available as an e-book.

Contents

List of Tables

List of Figures

Foreword

All progress in some way can be attributed to applying the right innovations in the right context.

While innovation is key, relevance is an equally key aspect.

In a country like India where there are a lot of problems, challenges, and opportunities, there is enough room for various kinds of innovations at all levels.

The problem is that India's focus on research, which forms the backbone for innovations, is tiny compared to the size of the problems, challenges, and opportunities in India.

Rather than taking a bystander approach, Varun has taken the approach of a curious detective and digs deep into the current gaps in India's science and technology ecosystem and provides a roadmap on how India can become a leader. Two years of research and hard work has resulted in *Leading Science and Technology: India Next?*

The topic of innovation, especially applied innovation, has been of interest to me, and I naturally gravitated to get more involved in what leads to that innovation—research and development (R&D). So I read this book with great interest.

To give you some context, the United States spends over $50 billion annually on basic and applied research. A major part of this funding goes to R&D centers in universities such as MIT. Research leads to creation of advanced technologies that will be relevant in the future. So part of the problem is solved. The other important part of the problem is to bring the innovations born out of this research to give birth to solutions that solve real-world problems.

When Jaishree and I noticed it, we decided to do something about it. So the MIT Deshpande Center was established in 2002 to extend relevant innovations born out of primary research at MIT to create meaningful impact in the real world.

It was clear that India does poorly in scientific research and innovation. But the question that was not clearly answered was "Why?"

Varun takes a data-based approach to answer this question objectively. He presents substantial new data and fresh insightful analysis. He also talked to more than 50 people involved in various facets of the research ecosystem. It gives voice to the key stakeholders—PhD students, professors, the industry, and administrators—on the strengths and weaknesses of our system.

The book covers a gamut of things—from making a case for why focusing on R&D is a must for India to stay ahead of the game to the key reasons why that sector is stalled (e.g., resources or lack thereof, accessibility, and merit).

Varun does not stop at pointing weaknesses, but ends the book with 16 well-thought-through principles for invigorating India's research ecosystem. You will find many interesting things including a few that will make you raise your eyebrows or frown; for example, one of the main limitations cited by researchers and PhD students was lack of enough travel money to attend conferences and meet collaborators!

One other finding that caught my attention was that there was a big scarcity of asking and pursuing answers for original questions. This is necessary to provide context for research—to connect it with challenging problems in India and the world.

Varun shares interesting real-world stories to make it a fast and interesting read.

Overall, I must say that someone had to write a book on this topic. It was long overdue. Varun took up that challenge and delivered his work in style.

Desh Deshpande
Founder of Deshpande Foundation and
Boston   Life Member at the MIT Corporation
September 2017   Author of *On Entrepreneurship and Impact*

.

Preface

We pride ourselves as having the third largest economy in the world. Since the liberalization of the early 1990s, businesses have grown spectacularly, creating great economic value for the nation. India has become the poster boy of the IT industry in particular. More recently, India has been recognized as the world's fastest growing economy, a silver lining on the rather dark cloud of recession and political crises elsewhere in the world. In 2016, India had the second largest number of start-ups in the world and was probably the fastest growing in sheer number of start-ups.

In stark contrast, we have the great challenges of poverty, hunger, and inequality in our society. As of 2013, 270 million people in India were living below the poverty line, accounting for more than 20 percent of the world's poor.[1] The Global Hunger Index 2016 ranks India as the 22nd most malnourished country.[2] India ranked 131 out of 188 countries on the Human Development Index, a composite indicator of human well-being, in 2016.[3]

We have a clear understanding of our human development problems. The news media discusses these issues often: national newspapers publish statewide statistics on various development indices and print stories of individual deprivation and hardship almost daily. Poverty has been a hot election item since the time Indira Gandhi was carried to the prime ministership by the "Garibi Hatao" campaign. Other than sops, the government has also introduced more serious and well-designed social development and economic redistribution programs. Aadhaar, NREGA, mid-day meal scheme, and Jan Dhan Yojana are all good examples. Recently, for-profit companies have started working to solve social issues in a sustainable way. These are called social enterprises. Many of these efforts show great promise.

In all, both intellectuals and popular media understand that India needs both economic growth and social inclusion/redistribution programs. Do we talk about scientific and technological innovation with equal zeal? No. These are neglected topics in India. Much of the discussion that exists is itself noise. We tend to discuss whether Ganesha's head was transplanted through technology thousands of years ago. The newspapers proclaim some useless so-called innovation produced domestically—usually a super cheap tablet, cell phone, or drone. The public marvels at such Indian achievements, which do not stand the test of scientific scrutiny. To add, a Nobel Laureate recently called the Indian Science Congress a "circus."[4]

There is some positive discourse as well. India's mission to Mars and our ability to launch satellites commercially are great achievements. One must admit, however, that we are doing things that others have done already, but at a lower cost. Our ingenuity is primarily our ability to do them at lower cost. Launching a person into space (never mind placing her on the moon) through a truly indigenous Indian program remains distant.

We also pride ourselves on *jugaad* and frugal innovation.[5] As an equal partner with frugality and limited resources, *jugaad* has enabled several successful innovations in India: Tata Nano, a low-cost refrigerator, a device for climbing trees, a lemon-peeling machine, and even a cotton-stripping machine that was recently invented by a farmer. With due respect for these achievements, they do not either comprise or substitute for scientific progress and basic research. *Jugaad* cannot provide either economic efficiency or social value to match the progress we have seen by way of genetic engineering, artificial intelligence (AI), and life-saving medical instrumentation.

Having spent more than a decade in developing research-led products, I felt compelled to write about the importance of science and technology for the future of Indian society and economy. India needs leadership in research and innovation to become a global superpower. The United States and China recognize the value of this leadership. They rank first and second in the world in research productivity. India ranks 12th.[6]

Research and innovation lead to new and differentiated products and highly valuable businesses. Products from Apple and Google are the result of several direct and indirect research efforts that took place in universities. Google's search algorithm was first developed in a doctoral thesis at Stanford University, while iPhone's proprietary touch screen, GPS, and Siri are the results of various government-funded projects and university-based research. Multiple companies across borders offer their services to build and sell new research-led products. However, the owner of the intellectual property (IP) claims the lion's share of the profit that these products generate. Apple, for example, takes the largest share of profits from the sale of Apple devices—the manufacturer receives just $10 per unit.

Research also leads to disruptions such as the Internet, the smartphone, and recent advances in AI. These create fresh challenges for incumbent businesses and a displacement of economic value to new businesses. India also will capture some of the economic value of these breakthroughs by providing relevant services, but it will be a much smaller portion than that received by countries where these disruptions happen. The largest Internet and mobile app companies are in the United States and China. These countries will host the most advanced AI companies also. Innovation elsewhere in the world has cost us in the past during the industrial revolution.

In addition to the economic value it creates, the pursuit of high-quality scientific research is a virtue in itself. Intellectual endeavor exercises the human mind and pushes our intelligence and creativity to its limits, creating new knowledge and new products to help us better understand the world and develop useful solutions to practical problems. We should never forget that high-quality research is responsible for all the advances in medicine that have helped us enjoy a better life.

Chapter 1 makes people aware of research: the place where it happens, who does it, what it achieves, and how it gets converted into economic value and public goods. Chapter 2 discusses why scientific and technological research should be a necessity than just a priority for India through six clear and compelling reasons. If we do not win the innovation race, we stand a high risk of losing the economic race as well.

The chapters ahead present India's current performance in research and how we can spur India to produce more high-quality research. I have also tried to dispel the usual stereotypes: Indians are superstitious, not creative, not curious enough, etc. Similar hypotheses are attributed to our slow rate of growth (called the Hindu rate of growth) which has been (thankfully!) debunked by a simple and straightforward liberalization of economic controls. Now we have one of the fastest growing economies in the world. Our creative potential in scientific research can be released by a similar straightforward plan of policy and action. Our problems are a matter of institutional structures, socioeconomics, and policy structures. I focus most of my arguments on these but reserve one chapter for an examination of the thought processes of our researchers.

I take a data-based approach throughout. My arguments are based on analyses of data from various public sources, universities, research databases, and surveys of faculty and PhD students. I have also talked extensively with university administrators, government officials, faculty members, industrialists, and PhD students to get a better understanding of how to interpret what I see in numbers.

Chapter 3 looks at a variety of metrics to measure the scientific progress of India compared to other nations. To measure the national output, we look at our "papers" and "citations" and at India's performance in doing high-quality and disruptive research in various fields in comparison with the United States and China. The United States and China outpace us by rates of 10 and 5 times, respectively.

I primarily cover research that happens in universities.[7] Across the world, most scientific inquiries occur at universities. I identify the critical building blocks that lead to high-quality research. Above all, the quality of research cannot exceed the quality of the researchers themselves. Chapter 4 examines the quality of our researcher cohort. We look at what deters our most promising young people from pursuing research careers, and we look at what we might change to make the field more attractive.

In Chapter 5, we look at resources needed by our researchers to do their work. These include such things as expensive brain-scanning machines and electrical and mechanical components. They also include more things such as travel support and competent technical and administrative staff. Our researchers are quite disadvantaged in all of these areas compared to our Western counterparts. We need to make resources accessible, enable acquisition of new resource fast, and allocate them according to merit.

Researchers need a work environment that is intellectually stimulating and supportive of their efforts. They need a wide community of talented colleagues and students who engage each other, affirm each other's discoveries, and help multiply their individual successes. A critical mass of researchers with the right incentives and opportunities to connect, collaborate, and compete is indispensable for our progress. This is discussed in Chapter 6.

In Chapter 7, we explore what it means to attain leadership status in research. Presently, a considerable portion of our research is derivative of the work done in the West. Since Independence, we have had few original or impactful scientific ideas. By building the right research culture, getting inspired by our daily problems, following better education methodologies, and developing scientific pride, we can ready ourselves to ask original questions and propose big, new, radical ideas.

To reap the benefits of our research efforts, we need to convert our research into economic value and public goods. Among MIT Technology Review's list of the world's "smartest" 50 companies from 2014–2016,[8] we find not one company from India.[9] Most of our companies are copycats of Western companies. In Chapter 8, there's a discussion on how we can develop an ecosystem for science entrepreneurship and promote interactions between research institutions and private industry.

In Chapter 9, we look at the big picture. What can the government do to help create world-class universities? How has government policy toward higher education and science evolved over the last 60 years? Do these policies support or inhibit research? What gaps exist in the university structures of today? What more can be done? India's future relies largely on how we answer these questions.

The last chapter proposes 16 principles for designing a well-functioning ecosystem as well as concrete steps India can take to become a leader in science and technology. These principles can be a guide for building better policies and also a yardstick to check their construct. I also throw in some creative and wild ideas!

This book is meant for two audiences: those who care about India's success and those who are concerned about the progress of humanity. I like to think that these two audiences are actually one. If this book instills in you any greater confidence in the power of science, if it motivates you to do your part to improve the Indian research ecosystem, and if it supplies you with the tools for it, then I will consider that I have succeeded in my goal.

I welcome comments, suggestions, criticism, and questions. Please e-mail me at varun.aggarwal@gmail.com.

Varun Aggarwal
Co-founder and CTO, Aspiring Minds

Acknowledgments

First, I would like to thank my parents. My father's Nehruvian dedication to science and education and my mother's Gandhian charity have molded me and my views. Without their guidance and hard work, I would not have pursued a scientific career and would have never known the marvels of research or ever have taken up writing a book such as this.

I started this book project in 2015. Many mentors, colleagues, and research associates have helped me on this journey. Professor Arun Sharma believed in the book as soon as I told him about it and gave me much valuable advice throughout the process. Professors Sanjay Sarma and Shailendra Mehta contributed substantially to the intellectual arguments made in the book. Both have thought deeply about the subject. I would also like to thank Mr Kris Gopalakrishnan, Mr Narayan Murthy, Dr Sumit Gulwani, and Mr Manish Gupta for providing their perspectives on the subject. Professor Henry Rosovsky was kind enough to meet me during my stay in Boston to share his insights and comments on my synopsis. I am indebted to Professors Arun Majumdar, Vasudeva Varma, B. N. Jain, M. S. Ananth, Ross Bassett, R. A. Mashelkar, Sanjoy Mitter, Kishore Trivedi (uncle), Philip Altbach, Ashutosh Sharma, Ramesh Hariharan, Rajesh Sundaresan, Lavanya Marla, Partha Pratim Talukdar, Pradeep Khosla, Shubhendu Bhasin, Jayant Haritsa, Ravi Iyer, and Rohit Karnik for taking the time to engage in extensive discussions on the topic. In addition, I owe thanks to the many faculty members and students in India and abroad—far too many to list—who provided their insights into how Indian research universities function.[1] My big thanks to all of them.

Several research associates helped me locate the data in this book and contributed analyses, tables, and graphs. They include Animesh Sharma, Ankit Rai, Pauranmasi Khurana, Mukul Agarwal, Abhishek Ahluwalia, and Palash Rastogi. I must thank Siddharth Nithyanand for his help at critical junctures of this project. Without their assistance, this project would not have been possible. I must also mention Shashank Srikant, Shubham Bansal, and Gaurav Gandhi for their helpful comments and for reading parts of the manuscript.

I am indebted, as always, to Professor Tarun Khanna for always being available to lend support and advice on the project. I thank profusely Professor Anirudh Krishna and Thomas Barlow, whose advice I sought from beginning to end. Dr Una-May O'Reilly, my advisor at MIT, continues to be a big support.

I thank the entire SAGE team for the help extended to me in bringing out this book. I will like to specially thank my editor, who was excited about the book from the word go. Her comments have been very helpful in carving the arguments in this book and making it tighter. I would also like to thank Robert Price for his proofreading of the manuscript, for providing helpful comments, and for rising to the occasion when needed.

But of all the people to whom I am indebted, a big thanks to my wife, Anna, who bore the brunt of the pressure of my writing while also bearing our son and bringing up a newborn. She never questioned the purpose of this book and has been a great source of emotional support and encouragement. She helped me greatly in deciding the final form of the book and in choosing the right publisher and marketing plan. Thanks so much (and sorry) to baby Naren. I stole his time to write this book—while he was in the womb and after. I hope he got inspired!

1

Science, Technology, and Innovation: An Introduction

By the time we get to the 2040s, we'll be able to multiply human intelligence a billionfold. That will be a profound change that's singular in nature. Computers are going to keep getting smaller and smaller. Ultimately, they will go inside our bodies and brains and make us healthier, make us smarter.
—Ray Kurzweil on the future of science and technology

A new kind of car is being tested on the roads of California. Once unimaginable, and recently only the domain of science fiction, this car is rapidly becoming a common sight: cars without drivers. You specify a destination and the car takes you there. It navigates the streets, maneuvers through traffic, obeys traffic laws, and watches out for pedestrians. Google built this amazing machine—the result of several significant advances in artificial intelligence (AI), global positioning systems, and sensors. Marvelous engineering synthesizes these components to form a viable product. You may spot such a car in Beijing as well: Baidu (China's Google) began testing its driverless technology in December 2015.[1]

If you look toward the sky—perhaps while traveling in one of these cars—you might see a drone, something like a toy helicopter flying on its own. If so, you are witnessing a similar experiment in autonomous piloting: compensating for varying wind speed, braving weather, and navigating around trees, buildings, and birds. Like cars that deliver us to our destination, these unmanned aerial vehicles (UAVs) can deliver our parcels through aerial routes, avoiding the slow and cumbersome process of traditional ground transportation. Why is this helicopter more remarkable than the toy helicopter your daughter flies at home? Precisely because your daughter isn't flying it—it flies on its own. It is an autonomous machine made possible by advanced control theory, mechanics, and data science.

Meanwhile, in a molecular biology lab in Europe, a scientist is growing a brain in a dish! Well, it is not really a brain just yet, but rather a three-dimensional bunch of neurons grown from a modified skin cell. For the first time, we can see how neurons grow and organize themselves into a three-dimensional structure. This is another important step in unravelling one of the greatest mysteries of our time: How does the brain work? The experiment could help answer another interesting question: How can we build the biological computer of the future, a computer with no electronic circuits, but neurons like the human brain?

From Europe, we cross the Atlantic to observe another biology lab, this time in Massachusetts, home to two of the top research universities in the world—MIT and Harvard. Scientists there are working on modifying our immune system cells to fight cancer. They edit the genes of these cells so that the cells can specifically target and kill tumor cells, without affecting healthy cells. Our immune cells do not possess this capability naturally, and if we wait for evolution, we might be waiting for another thousand years. No need! Human knowledge, creativity, and passion are making it happen today to enable new methods of treatment. Doctors can already take immune cells from the cancer patient, genetically modify them, and push them back to fight the tumor. The next step is the creation of a bank of generic cancer-fighting immune cells, which could be injected into any cancer patient—very much like a blood transfusion. At least half a dozen start-ups are working on various competing and complementary technologies to make this a reality.

A professor at the Chinese University of Hong Kong (CUHK) is fighting cancer in a different way—by detecting it early. Scientists in the United Kingdom estimate that one in four cancer patients lose their lives because of late diagnoses.[2] Our CUHK professor is busy sequencing the genes in a drop of the patient's blood to detect the cancer early, by detecting the DNA that cancer cells shed into the bloodstream at the time of formation. Scientists sequenced the full human genome in 2003: it took 13 years and at least half a billion dollars. Today, scientists can sequence an individual's genome in less than 26 hours and for less than $2000. This cancer-detection technology is still in the lab and is not ready just yet. But believe me, it will arrive sooner than we might think: as quickly as the staggering speed with which the gene-sequencing technology was developed.

These examples state the obvious—we are living in exciting times of huge advances in science and technology. Our knowledge and methods have become more mature over time, and we are making stunning progress in solving real problems. The impact of science and technology on our lives is closer and larger than it has ever been. Technology is creating public goods through advanced disease detection and diagnosis methods, newer kinds of drugs, and surgical devices and methods. Moreover, by way of mobile devices such as smartphones and iPads that place technology in each person's hands, people are more connected to markets, social benefits, and employment opportunities.

Science and technology impact business in a big way. Technology has led to the creation of new economic value and also to the disruption of the older ways. Consider the iPhone: a disrupter of older industries such as cameras and music devices and a substrate for several new apps and devices businesses. The iPhone carries sophisticated AI algorithms, the power of global positioning systems, and advances in touch screen interfaces. Similarly, our seamless experience on Twitter, Google, and Amazon carries advanced cloud-computing architectures, smart data management systems, and sophisticated machine-learning algorithms. Amazon, for example, recommends new books to you by utilizing advanced machine-learning algorithms and processing large amounts of data on what you and others previously purchased.

These are just the things that we can see. In fact, science and technology permeates our lives in ways that we do not even recognize.

How Does Innovation[3] Happen?

How did we arrive at this fantastical point in time? Google's driverless car, the iPhone, and Amazon's fleet of helicopter delivery bees might make you think that entrepreneurs and the start-up ecosystem are at the heart of innovation. You would not be entirely incorrect. Start-ups and entrepreneurs are very important to the innovation ecosystem. However, at the heart of innovation is scientific research. Innovation begins in the universities with research projects that lead to research findings. Innovative entrepreneurs then exploit these findings, combining multiple technologies and smart engineering to develop useful products and bring them to market.

Again, consider the iPhone. Mariana Mazzucato, a professor in the Economics of Innovation, tells us that most of the critical technologies in the iPhone were created through university[4] research and government-funded products.[5] Most of this happened in the United States. The multitouch scrolling and gesturing technology for screens was developed by Westerman and Elias at the University of Delaware. The research that culminated in this technology was developed over decades of study into capacitive touch screens and was first initiated in the 1960s by a British government agency. The GPS systems that enable our maps and navigation were built by the US Department of Defense (DoD) for military purposes in the 1970s (the technology was not released for public use until the mid-1990s). The speech recognition that Apple's Siri uses to understand your instructions has its roots in the work done at Carnegie Mellon University (CMU), in the lab of Raj Reddy, a scientist of Indian origin. And speech recognition technology is already being disrupted by a new technique deep learning, invented at the University of Toronto.

Similarly, research in driverless cars were first spurred by Grand Challenges funded by Defense Advanced Research Projects Agency (DARPA), US' defense research agency. In the 2005 competition, 22 cars covered more than 11.78 km of distance in a desert with narrow tunnels and sharp turns. Stanford University's car covered the distance in the least amount of time, followed by two cars from CMU. The 2007 challenge had an urban course, this time won by the team from CMU followed by Stanford University. Not to mention, the cars utilized several research advances in machine learning and robotics, which took place in universities over the years.

Jonathan Cole, the ex-provost of Columbia University, has compiled a list at www.university-discoveries.com. Here, a reader can see the myriad discoveries and inventions that have their roots in US universities. The list will blow your mind. It includes most of our basic means of entertainment and their first avatars, such as the black and white television, the FM radio band, and movies with soundtracks. The list also includes light emitting diodes (LEDs), the autonomous vehicles previously mentioned, helicopters, and even chlorinated water purification systems. The discovery of vitamins E and K happened at the University of California, which also discovered and identified the HIV virus as the cause of AIDS. The University of California (San Francisco) also gave us the cochlear implant that is helping deaf people to hear. Magnetic resonance imaging (MRI), used routinely for medical diagnoses, came from Stanford University. Antibiotics were first developed at Rutgers.

Universities are commonly known for teaching. So how do they find the time and resources to discover and create all of these wonderful things? As repositories of learning, universities disseminate knowledge. However, they also create it. Professors teach by lecturing, but they also lead investigative endeavors into open problems, develop new theories, and establish new facts. They use systematic scientific principles in this pursuit, while also discovering new ones. This process is called research. Professors conduct research groups, usually comprised of students pursuing their doctoral degrees. The professor and the students take up challenging, previously unsolvable problems, and then create new knowledge to solve them. They announce the results of their research in papers that are presented at conferences and published in journals. Volume of research output and the number of opportunities for doctoral-level research work are key metrics in determining a university's rank. The top universities consider producing new research results and training new researchers through their doctoral programs, a key responsibility.[6]

Funding for university research comes from a variety of resources. Federal governments are a prime sponsor of scientific inquiry. For instance, government funding agencies include DARPA and the National Science Foundation (NSF) in the United States, and Department of Science and Technology (DST) and Department of Biotechnology (DBT), among others, in India. Large corporations fund university research as a way to get universities to concentrate on their problems of interest so that they can use the results of the research for commercial application. In most cases, the results of research are a pubic good regardless of the source of funding—the results are openly disseminated and can be used by anyone and everyone.[7]

At the next step, innovative entrepreneurs bring the fruits of these research projects to the market and into our lives. Quite the contrary, these entrepreneurs are anything but technologically poor, business-minded people who just take knowledge from the universities and capitalize on it. Conducting research is one thing. Creating useful products and services with wide application is another. In fact, this step requires considerable innovation and is led by the entrepreneur. Once again, consider the iPhone. Everyone had access to the technologies involved. But we know of only one person—Steve Jobs—who conceived of such a product and thought it possible, thereby disrupting the mobile phone market forever. Putting all these technologies together in a compact, usable product amenable to mass production is a challenging task to say the least. It requires much additional research than merely that which a scattering of university projects provides, as well as a ton of smart engineering. Private industry and entrepreneurs do this step—conceptualize products and services and then actually build them.

The same is true with the driverless car. When it was first built for the DARPA Grand Challenge, all the enabling technologies already existed. This is known as recombinant innovation: smartly combining existing technology to build innovative products, the like of which have not been seen before. Industry does a lot of this. We do it at Aspiring Minds. We created software to automatically grade free speech. The software asks you to speak on a topic and automatically assigns grades on your pronunciation and fluency. Our technology involves a speech recognition engine, crowdsourcing, state machines, and supervised machine learning. All these technologies previously existed, yet an automatic free-speech grader did not. The work that went into developing this was recognized at a top computational linguistics conference, qualifying it as research in itself. The product itself is greatly useful and is employed by several of our customers.

This is the way in which scientific advances enter our lives. Universities and entrepreneurs work in a "conspiracy." The universities create new scientific and technical knowledge, while the entrepreneur exploits this knowledge to create new products and services. In the process, we see the creation of economic value, new businesses, and public goods. The majority of university research is publicly funded, while the profits go to the entrepreneurs. The profit is turned back into social advantage and public wealth through economic expansion, jobs, taxes, philanthropy, a more enriched intellectual and cultural life, and of course, further investment in research.

In describing the process of innovation, I have taken the liberty of oversimplification. The actual real-life process of innovation is much more complex. Let us look at some important nuances. Research happens not only in university laboratories but also in government-owned labs and private labs. Government operations include NASA for space research, the DoD for defense research in the United States, and the Defence Research and Development Organisation (DRDO) and Council of Scientific & Industrial Research (CSIR) in India. Microsoft Research, IBM Research, and historically the Bell Labs are all examples of private research laboratories. Many scientists in these labs work on fundamental research as well as recombinant innovation. Bell Labs has introduced many scientific disruptions including an efficient algorithm for linear optimization in 1984 by Narendra Karmarkar, another scientist of Indian origin. In 2011, IBM Research built Watson, a computer that could answer questions posed in natural language. In a remarkable demonstration, Watson defeated several human contestants in the game of Jeopardy.

Presently, the involvement of industry in research is at an all-time high. Facebook, LinkedIn, Baidu, Google, and companies like my own regularly publish papers in data science conferences and other forums. The speed of innovation is very fast, and many technologies have reached a degree of maturity whereby they are immediately applicable in the real world. Increasingly, market survival practically requires industry to delve into research to beat out their competitors in developing new products. I reckon that private industry busies itself with things that can be commercialized in 5, if not 10 years. Universities, on the other hand, work on problems where progress might not be realizable for 10–50 years. Universities have been working on AI since the 1950s, while the breakthrough for the technology to be useful only came in the early 1990s. The new frontier is brain research. Universities are taking the lead here as well.

While industry trespasses on the territory of universities by doing their own research, universities strike back at industry by building products, owning patents, and incubating start-ups. MIT Media Labs was one of the early identifiers of the research and innovation aspects of building products. Some of us at the Computer Science and AI Lab at MIT scoffed at their work, not considering it to be real research. We didn't laugh long. Great innovations have come out of Media Labs such as Scratch, a programming language for kids; LEGO Mindstorms, a kit for kids to build robots; and a hundred-dollar laptop. Furthermore, top research universities have a systems approach, whereby they build large systems to further their research. Advanced Research Projects Agency Network (ARPANET), the earlier avatar of the Internet, is an example. So are several robots built by Rodney Brooks' group at MIT. Building these systems require collaboration among multiple researchers and many a time from multiple disciplines. Building smart robots is a classic example that brings together computer scientists and electrical and mechanical engineers together. Today, multidisciplinary research is in vogue, where collaboration goes beyond disciplines within engineering, to health and social sciences.

Universities, professors, and students retain part of the economic value of their innovations through patents, consulting work, and through resulting awards of more research funding. They may also obtain a stake in the companies using their technology. Professors sometimes take sabbaticals to become entrepreneurs, and many PhD students take up full-time entrepreneurship. Google, which was started by two Stanford PhD students, Sergey Brin and Larry Page, is a classic example.

Innovation happens through the efforts and interactions of research universities and private industry. The first and necessary conditions are the courage to attempt to solve vexing problems, the willingness to invest in long-term projects, and dedication to producing high-quality research. Together with this, a buzzing innovation ecosystem translates research results into innovative products and services.

As you read this you might be experiencing the fruits of years of research and innovation efforts—Kindle, an ultralight and power-efficient device, to read books.

The Research Ecosystem

Scientific innovation requires a precursor: scientific research. Research is mostly funded by public money. One might argue that research results are public goods, no matter the country in which they appear. Results are published in open papers and journals, and anyone can take advantage of them to build innovative products and services.[8] Why not leave the expensive pursuit of research to the West and then simply partake of the harvest given our sparse public resources? While it might sound practical, reality does not work this way. Innovation does not occur unless researchers, entrepreneurs, and investors work in a common ecosystem in close proximity. It is most likely to occur in the place where scientific research happens. The people who are most likely to take a professor's research to market are the people closest to him or her: colleagues, students, investors, and industry connections. They have the most intimate knowledge of research efforts and are aware of the results even before they are published. Replicating the results of a research project, however, is easier said than done. And of course, the difficulty will vary by field. A robotics project is more difficult to replicate than a math study. A primary investigator or a first witness can lap up groundwork laid out in the university rather than spending 6–12 months trying to replicate just the first set of results. The proximity to the place of research readily makes available entrepreneurs, manpower, and technical know-how and provides a first mover's advantage for a business.

Equally significant, the mere presence of such a research eco-system around the university provides the motivation and confidence required for innovation. Cutting edge research might seem glamorous, but it is risky by nature. An individual needs a lot of encouragement and support to move away from safe efforts and engage in areas where success is not guaranteed. Most experiments fail to produce a robust product. Most often, they fail to provide anything useful to the market at all. The few experiments that succeed create huge business value. Life experience provides the motivation to try: seeing others attempting and succeeding in similar experiments in an active and supportive ecosystem. There must be a critical mass of entrepreneurs to generate success that motivates people. At the same time, there is a need to have respect for failure. A tradition of failed experiments, due to lack of critical mass, leads to expectations of more failure and a disinclination to try. And then, investors must be present who are ready to invest in risky businesses with long gestation periods.

There is considerable empirical evidence around geographical clustering of research and innovation. Research pioneered by Maryann Feldman, a professor of public policy at University of North Carolina, shows that location and proximity strongly influence exploiting knowledge spillovers.[9] Examples include MIT and Stanford, and the areas surrounding them. Companies started by Stanford community in the Silicon Valley generate $261.2 billion revenue, 55 percent of the total revenue generated by "Silicon Valley 150" companies. Massachusetts had 1,065 MIT-related companies constituting 5 percent of total state employment and 10 percent of the state's economic base. MIT-related firms accounted for 33 percent of software sales in the state and 25 percent of all manufacturing sales.[10]

The industrial sector does not merely take from the research community. Research is not a one-way pipeline. The financial and industrial sectors inform research in turn. They signal their needs and unsolved problems to the research community and offer up these problems as questions awaiting solutions. The ensuing discussion between the two communities generates new ideas, which industry and research then pursue in tandem. Hence, we close the feedback loop of innovation.

This is how progress in science, technology, and innovation happens!

2

Leading Science and Technology: A "Must-do" for India

In a world where the powers are determined by the share of the world's knowledge, reflected by patents, papers and so on ... it is important for India to put all her acts together to become a continuous innovator and creator of science and technology-intensive products.

—A. P. J. Abdul Kalam

India aspires to be a global leader. If we accept this as true, then we must also accept that India must lead science and technology. Such leadership comes through the pursuit of great scientific research. Research creates new knowledge and advances science and technology. Alongside economic growth and social development, I posit that scientific research is a national necessity. I submit six compelling reasons why it must be so.

1. We Lose If We Don't Win

The next best provider might be as good, but it will not matter.

—Erik Brynjolfsson and Andrew McAfee, *The Second Machine Age*

Consider the task of identifying human figures in surveillance video. The task will generate considerable work for multiple businesses, given the sheer number of workers required to annotate so many million images. What if an algorithm is invented to perform the work automatically? Once invented, the cost of replicating this algorithm is quite low. Then, distribution through the Internet will allow the algorithm to circulate widely around the world for practically no additional cost. In time, the algorithm will become better and better as it is exposed to more data accrued in the form of images from its ever-increasing number of users.

However, this algorithm will not be the only one. Others will follow. In the shakeout, one will become the best. But what will be the fate of the second best algorithm? Will it find a fan base, or will it be of no use at all? In a traditional market, this second best product might still find an audience, because of limitations in market access. However, the Internet has broken such barriers to access.[1] Over time, the second best algorithm will attract fewer and fewer users, resulting in less and less exposure to data and therefore less and less opportunity to become smarter. The gap will widen quickly and at an ever-accelerating pace. Eventually, the inferior algorithm will simply atrophy and lose.

You might be surprised to learn that computers have already mastered the task of identifying human images. Facebook has now solved an even more difficult problem in tagging faces with names automatically!

The world is presently engaged in a "winner-takes-all" market for products and services. We may attribute this to three factors. One is the digitization of products and services, and the human intelligence that makes it possible. Fast replication at little to no cost. The second factor is the ease of access and distribution made possible by the Internet and various other communication channels. And the third is called the "network and accumulation effect," in which a product or service attracts more users by virtue of already having a large number of users. A well-known example of this "winner-takes-all" situation is Facebook. It is practically the only social network that we all use today. Similarly, most of us across the world use the same search engine, the same chat messaging app and the same micro-messaging website.

In today's market, we lose if we do not win. Therefore, our businesses must aspire to be more than merely good or the second best. They must be the best if they hope to grow and derive economic benefits from their respective sectors. Innovation must return to center stage. The products and services that are poised for success are those that are backed by cutting-edge research and technology. Innovation, innovation, innovation! We need to continue to innovate to be the best and continue to be in business.

If we do not innovate, then Western leaders of innovation will simply overwhelm our slow performers, and we will become like outsiders looking in. For this reason at least, I argue that research and innovation is not merely a priority, but an absolute necessity. Uber sees fit to develop its own Google-like maps and its own driverless cars. Is Ola following suit, or even thinking about it? Pardon me if I am wrong, but I think not.

And lest we become discouraged, we must remember a fundamental point: the winners can come from anywhere in the world. There is no reason why WhatsApp could not have been an Indian company. Market access does not depend on being in a New York, Silicon Valley, or London. Modern companies transacting in digital good do not require local production units nor are they bothered by high transport costs. If Kanpur enjoyed an innovation ecosystem around its IIT, there is no reason to doubt that our Kanpurian companies could be global winners. In this new world, anyone can be a winner and anyone can reap the benefits of their efforts. Yet still, no person succeeds alone. We need a local innovation ecosystem to breed champions. Furthermore, the "winner-takes-all" system is leading to the slicing and dicing of products and services into smaller units, which makes room for a large number of innovators (see "Many Winners!"). We can win in some of these categories.

Many Winners!

The "winner-takes-all" world is leading to the slicing and dicing of products and services into smaller units, which makes room for a large number of innovators.

Because of the simplicity of replication, low sales and delivery costs, and access to the entire worldwide market, narrow products can sustain relatively large businesses. Although the cost of research and development of these very specific products might be high, the gains from selling to the large global market make it possible to recover the investment. For instance, if one can make an extremely good surveying tool, just a surveying tool, there can be substantial revenue selling it to the huge global market. Surveymonkey.com with an annual revenue of more than $100 million is an example. In earlier times, these innovations did not happen, due to their low economic value because of limited access to the marketplace.

Today we have separate companies for the best photo management tool, best chatting tool, the best referral management system, and the best e-mail system. Even large companies integrate narrow products and services from other entrepreneurs in their products—from those who do these components really well. There is an opportunity for a smart innovator to take the lead in multiple narrow product areas. This creates room for a larger number of winners and opportunities, but each needs to be world class!

In order to not lose, we need to win. Moreover, we need to win today. If India does not get scientific research-led innovation right, our businesses will steadily, and surely, and not so slowly lose the global race. It is not a priority but a necessity, if we are to keep the India story alive.

2. Leapfrogging Economic Growth

These "iProducts" are all assembled in China and contribute to the US–China trade deficit to the tune of $150 per unit for the iPod, $229 per unit for the iPhone, and $275 per unit for the iPad. Yet the value captured in China from the sale of *these devices is estimated at only around $10 per unit, with some returns flowing to component-makers in Japan, Taiwan and South Korea, but with by far the largest returns actually retained by Apple.*

—Thomas Barlow, *Between the Eagle and the Dragon*

Products and services that cannot easily be copied or produced for a lower or competing price command the highest price in the market. Now as always, maximum economic value is captured by differentiated and irreplaceable products and services. Not surprisingly, such products and services are usually the result of research and innovation.[2] For example, even though Chinese companies make money manufacturing Apple products, Apple captures 5–10 times more value by unit as the Chinese manufacturing plants. Apple owns the innovation and intellectual property (IP). In the building of the iPhone, the Apple company is much more difficult (if not impossible) to replace than any subsidiary link in the manufacturing chain. Now as always, the innovator captures the largest share of the value created.

India's current stable of IT services, business process outsourcing, and manufacturing companies add value to our economy. They offer trained manpower and efficient processes for the completion of tasks. Our companies are not unique; others can provide similar services. Our advantage lies with cost. We couple availability of cheap, trained workforce with efficient business processes.[3] On the other hand, innovation-led companies can create substantially more value and can create it faster, by virtue of their products' irreplaceability and perceived value among users. More often than not, innovation-led companies are product companies. The innovation is packaged as a differentiated product with a built-in fan base. Microsoft Windows, MS Office, Adobe Acrobat, Apple's iPhone, and Google's search algorithms are much more difficult to replace than the services provided by Infosys, Wipro, or Genpact, for example. The aforementioned companies are much larger, they grew astonishingly more quickly, and their economic value per employee is much higher. The innovation that they offer was only possible through exploitation of research results. India needs to create such highly valuable companies to grow faster.

Also, research and innovation leads to widespread disruption on a macro scale. Probably, the first such disruption was the Industrial Revolution, when mankind first learned how to automate human effort. This allowed us to build and replicate goods at a much faster rate and at a lower price than what individual craftsmen could do earlier. This mechanization wiped out the traditional industry of handmade goods. Prior to the Industrial Revolution, India produced 23 percent of the world's gross domestic product (GDP). Today it is around 7 percent. For various reasons that are all hotly debated, India could not participate in this change or share in the benefits. Not being the authors of the innovation and India's lack of geographical proximity to innovation are two important reasons why we fell behind.

The next major disruption occurred in the 1990s with the introduction of the Internet. And one might argue that we experienced a third equally significant disruption with the sudden pervasiveness of the smartphone. These developments completely changed how we access information, goods, and services. It created a substrate for several new kinds of products such as search engines, e-mail clients, social networking, and e-commerce platforms. In addition, extant non-Internet based services were replaced. The mobile phone again expanded the market to a much larger consumer base and provided avenues for even newer services through the use of GPS and the touch screen. All these impressive macro-disruptions happened in the West, buoyed by research in universities plus the entrepreneurs capable of bringing them to market. The largest Internet and mobile app companies, in terms of economic value created, are located in the West and some in China.

As we debate, another disruption is taking place: the mechanization of human intelligence. Recent advances in AI boggle the mind. We have already discussed the examples of driverless cars, face identification in images, and intelligent drones. But this is already yesterday's news. Coming up the pipe are automatic image recognition, chatbots, automated video interviewing, and speech recognition. Each of these has been buoyed by research, mostly in universities and sometimes in industry. The next couple of decades will continue to see several disruptions in 3-D printing, Internet of things and our understanding of the brain, and the genome. But, as usual, these disruptions are all coming from the West—meaning that the West will again claim the lion's share of the economic value created.

Advances in AI will disrupt older business models on a scale that is yet difficult to imagine. AI technology will render much human input obsolete. This will take away business from our IT and business process outsourcing companies. Our price advantage in cheap labor will shrink. This constitutes a new and rather imminent threat to business in India.[4] Similarly, combining AI with advances in robotics will disrupt manufacturing. Factories will have more robots than people. South Korea has 478 robots per 10,000 workers, the United States 164, Japan 315, and Germany 292.[5] China, at 36, is investing hundreds of billions of dollars in automation.

Macro-disruptions, sparked by research and technological advances, displace economic value from old players to new players. It also creates opportunity for newer kinds of products and services. The innovation ecosystem and researchers' proximity to it are the key essentials for capturing a share of the new economic value. These essential components of research can propel the creation of economic value for a company or a country, allowing a nation to leapfrog to a leading position.

Lastly, there is considerable academic evidence for a connection between research productivity, innovation, and economic growth. Some of these studies include India and answer how scientific research influenced economic growth in the past (see "Scientific Research and Economic Growth").

Scientific Research and Economic Growth

In the last two decades, various studies have examined if and how research publications influence innovation and economic growth. Various input parameters such as number of research publications, research expenditure, and research impact (and their per capita values) are used as proxies to investigate output and quality. The measures of innovation, business impact, and economic growth include number of patents, GDP, and survey data from industry. The studies examine such data from various countries across various time periods.

Do a higher number of research publications in a country lead to higher per capita GDP? Studies differ in the statistical methods they use to establish such a relation. A considerable number of these studies have found a correlation between per capita GDP and per capita research productivity.[6] This implies that during the years, a country had higher research productivity, it also had higher GDP, controlling for all other factors. Of course, this does not necessarily mean that the higher research productivity is the direct cause of the higher GDP. Instead, the correlation might be in reverse: higher GDP drives higher research productivity by creating more funds for research. Another possibility is that an unknown third factor, such as population decline, is causing the increase in both GDP and research productivity per capita.

Studies have tried to identify which of these relationships, if any, is actually true. There is differing evidence due to the various methods and statistical techniques that each study uses. A 2016 study by Solarin and Yen examined 169 countries including India for the period 1996–2003.[7] They find that research output in both developing and developed countries has positive impact on economic growth. The paper finds that the impact is larger for developed countries compared to developing countries, which, we may speculate, is due to better and more mature innovation ecosystems. In their conclusion, they write, "the policy implications of these findings are that more R&D programmes are needed to ensure more research output especially in the developing countries with very low research activities."

Another 2016 study by Kumar et al. finds that research publications per worker positively influences output per worker in both the United States and China.[8] For China, research and output have a mutually reinforcing effect. In their 2013 paper, Jin and Jin observed that research publications have positive and significant effects on economic growth in 46 countries studied, including India.[9] Inglesi-Lotz et al. examine the causal relationship between GDP growth and research output in the BRICS countries from 1981–2011.[10] Interestingly, they find a bidirectional causation between growth and research output in India, but in no other country.[11] This suggests that India, like China, already has in place some of the innovation ecosystem infrastructure to not only derive economic benefits from research but also to use these economic benefits to further research.

3. Better and Less Expensive Public Goods for the World

India's Mars orbiter, the Mangalyaan, cost $74 million. This price tag is about 10 times less than the US' Maven orbiter. How did we do it? We did it by carrying a lighter load, true, but we also did it by utilizing our traditional advantages in cheaper indigenous components, lower manpower costs, and so on. Mashelkar has often mentioned that India has the highest number of journal papers per GDP per capita.[12]

As with the Mars orbiter, India has the potential to perform research at a much lower cost than the West. This can be our competitive advantage. Lower costs can help attract research investment to India, just as our lower costs have attracted investment in the services industry.

Research as a public good helps mankind make progress. The ability to do research cost effectively is a boon not only to India's bottom line but to mankind entirely. For example, if ISRO were to develop and produce medical instrumentation and equipment with the same acumen that it used to build the Mangalyaan, it could provide the world with substantially less expensive scanning and diagnosis machines. Given our own low performance on various public health indices, India should make this a national purpose in good conscience. Our leadership role in the developing world makes it a responsibility. We can assist the entire world in public health and human development issues, while creating economic value. This is much better than simply handing out millions of dollars in aid.

There are a couple of pitfalls to consider. Currently, India does not do high impact research. Our research investment is low, so accordingly most of our research is of low-to-moderate impact. One might argue that this is what leads to low cost of research.[13] While I agree that some of India's cost advantage is accredited to these phenomena, I am not saying that India concentrate on mediocre research for the price advantage. Rather I say that as India advances toward high impact research, we will retain a share of our cost advantage. We might not be 10 times cheaper like now, but we will still be 3–5 times cheaper. This is where we should aim: research quality comparable with the West, but at a 3–5 times lower cost. That will make us hugely competitive.

Another pitfall is that our low-cost innovation results in products and research that are themselves of substantially lower quality. For example, we are proud of our supercheap tablets at 20 percent the cost of a Samsung tablet. Too bad, it is not even 20 percent as good as the Samsung! Focusing on such trinkets is detrimental and mis-leading for Indian research. We need to save ourselves from such rhetorical claims and ensure that our innovations can withstand the highest scientific scrutiny and stand side by side with their global competitors.

We do have some better examples of course. The Jaipur foot is a prosthetic device for people with leg amputations. Functionally, it cannot compete with the best prosthetic devices available. But it does provide a good alternative for people of limited means and in certain environments. The Jaipur foot is a prime example of how India can do things right, using smart engineering, cheap com-ponents and manpower, and a few clever hacks. For products of narrow application and limited functionality, with consideration for affordability, we might say that India performs rather well. Remember the Tata Nano. Nevertheless, these cannot comprise the whole of India's research. Such innovations are a respectable part of our innovation system, but we simply cannot allow them to substitute for the progress that basic research provides.

In contrast to these, there is also research whose goal is to reproduce an existing functionality, without any limitations, at a lower cost. This might include a new method for detecting diabetes or an eye condition. For example, Ramesh Raskar, an Indian-born scientist at MIT Media Labs, has found a way to do cheap and scalable eye testing and glass prescription. The accuracy remains the same, but the cost is lower. These savings are not small. In fact, the savings are usually exponential. Such innovations are examples of great research, and they offer much more value to the world compared to things like our Indian tablet.

India can be a leader in great research at low cost. To get there, we need multiple approaches. We should focus on doing what no one has done before: for example, finding a cure for cancer. If we can achieve such feats, then we will be doubly advantaged because the cost of the resulting product will be naturally lower than if the same product had been developed in the West. Next, we can add to our fame by developing lower cost techniques for existing services such as eye exams and water treatment. This benefits ourselves and the world. Then, we should utilize our traditions of smart engineering, cheap domestically produced components, and cheap manpower to produce affordable alternatives for prohibitively expensive but necessary goods. All these things together, when smartly balanced, comprise India's competitive advantage as well as our moral responsibility to ourselves and the world.

4. With Great Talent Comes Great Responsibility

Indians outside India have done tremendously well in scientific research. Several statistics illustrate this. For example, 12 percent of all scientists in the United States and 36 percent of all scientists at NASA are of Indian origin.[14] Indian citizens currently comprise the second largest international student population in the United States after China[15] and are rapidly gaining ground. Until recently the head of the NSF—the apex research grant-making body in the United States—was Subra Suresh. The current president of The Royal Society of London—one of the world's oldest independent scientific academy—is Venkatraman Ramakrishnan, a person of Indian origin. The success of Indians in research and technology in the United States is a plain fact.

Before Independence, some of our most noted and influential scientists actually did their research in India. One of my personal favorites is Jagadish Chandra Bose,[16] who did pioneering work that contributed to the invention of radio and to our understanding of plant nervous systems. Other notables include C. V. Raman and his pioneering work in physics; Srinivasa Ramanujan in mathematics; and Satyendra Nath Bose and Meghnad Saha, whose contributions disrupted world science. This was a time when we did not lose the bulk of our talent to other countries. Indians managed to do great work despite a fledgling modern education system and limited resources.

Our traditions of science and math go back further. Some of our earliest scientists made great contributions to the fields of geometry, trigonometry, and algebra. I recall that my math textbox in school had a shloka by Bhaskaracharya (1114–1185) on how to find the roots of a quadratic equation. Panini in the fourth century BC systemized Sanskrit grammar and is considered the father of modern linguistics. Indians of past times made great contributions to astronomy and metallurgy.

Unfortunately, current discussions of India's contributions to science and technology are laden with unsubstantiated assumptions, rhetoric, and arguments sans evidence. Such discourse disrespects our scientific tradition and degrades it. I see a need for a proper evidence-driven study of India's contributions to science and technology. I strongly believe that such a study would inspire much greater respect for what Indians have done. It would inspire our countrymen and possibly generate some fresh scientific insights. Such work would require the collaboration of historians and scientists across fields.

Indians have always shown great aptitude for science. Before Independence, Indians did great things in India. Later they departed in order to do great things in the West. Our people have a natural interest and ability to do great science. Therefore, any investment in research will have great return on investment (ROI) given that top talent is available and ready.

Let us be aware that we have a huge underutilized talent pool that can do great research. Our failing is that we cannot seem to attract these talented and capable people to careers in science and research. For those that we attract, we cannot provide the types of environment that are conducive to success. In India, only 0.23 percent of people with college education do research, whereas around 1 percent of them do research in China and the United States.[17] This is a huge loss of opportunity. Great talent, if nurtured and guided correctly, can be the creator of new knowledge, which will not only contribute toward the Indian economy but also help us solve our problems in social development. And again, as we solve our own problems, we will better position ourselves to help others.

And amid all this talk of responsibility to economy and mankind, let us not ignore the responsibility we bear for our talented young minds: the minds of our young people, but also the young minds inside each of our own heads, regardless of our age. Pursuing the goal of high-quality research is a virtue in itself. It exercises the human mind and pushes it to its limits, wherein we realize how to begin to dream of what new knowledge—and truly new innovation—looks like. If we are not consciously working to provide opportunities and environment for Indians to realize and exercise their capacity, then we are betraying the Ramanujans and Jagadish Boses, and all those gifted with scientific minds to reach their true potential and contribute to the world.

5. Science Must Be the Basis of Our National Development

Science must be the basis of our national development. Our societal goals arise from our morality, ethics, and culture. Science provides our tools and methods for achieving these goals.[18]

The scientific approach depends itself on "evidence" for the validity of any statement or method. Scientific research is an endeavor to find evidence or proofs to establish the truth of any given set of statements. It also continuously searches for new ways to address our questions, generating new statements/methods in the process, but relies consistently on evidence for veracity. It is this exploration that leads to new inventions and discoveries. To many, this evidence-based approach is known as "rational thinking."

As our society progresses, we will confront many questions: Is it useful to massage a baby's body? How can we clean the Ganga? How do we educate our millions of young kids effectively? At what age should we provide sex education, if at all?

We have no shortage of people to answer such questions. Many respected individuals—politicians, bureaucrats, business people, news anchors, journalists, and the *panditjee* in the mandir—come forward to offer their thoughts on our pressing national questions. These are great guys and gals with great ideas, and their suggestions are welcome, but their response needs to stand the test of scientific scrutiny.[19]

More importantly, we need to hear from the experts—the *vidvaan*—who consider these questions from a scientific viewpoint. These are the folks who know answers backed by strong scientific evidence. They know what constitutes defensible evidence and what doesn't. They are great voices to hear on our questions of national importance and great commentators on the opinions of others. Furthermore, they are the best candidates for breaking the cocoon of established thinking and bring new ideas through the application of scientific process. Relying on these experts will give us a better chance for progress than if we continue to rely on our traditional authorities. We need a great number of scientific women and men, and we need them to participate in our public discourse.

Today we see the rise of irrational elements worldwide that view science with suspicion and even mockery. We see rationality displaced by jingoism. In India, superstition abounds on TV, which presents a revolving circus of magicians, ghosts, and man-beasts. Even news programs prefer to recruit paid and self-interested panelists who speak with little scientific basis, but also deride the scientific method. A country with a low literacy rate such as India can be ruined by trafficking in such spectacles. This is far from Vivekananda's and Nehru's vision of an India devoid of superstition.

Science might be our only hope today. Great scientific research leads to the spillover to create a scientific society. A scientific society and great scientific research present a symbiosis, each strengthening the other leading to a powerful feedback loop. Great universities form the centers of scientific thought, which, in turn, inspire and influence society. Great researchers influence society in many ways: as guides, teachers, parents, advisors to business and government, and as public intellectuals. They are society's repositories of knowledge, aware of the newest methods and findings. They transmit their learning and thought processes to those they touch in the course of life.

To realize the benefits of science in our society, we need a critical mass of great researchers in India. To compete with the loud voices of the scientific naysayers and doubters, we must amplify the message of the science through sheer numbers. We also need the right institutional structures to connect our scientists with the society, through ways of media appearances, newspaper columns, and books.

Science is the most important ingredient: a *humsafar* (co-traveler) in India's progress.

6. Now or Never

Now is the most opportune time for India to reinvigorate its scientific research agenda. By far, the most important reason is that people are finally ready. Science and technology are no longer the domain of the few, but have touched the lives of the masses. People have seen how the IT services industry and the mobile-based economy have provided jobs and made a number of daily tasks easier. The fruits of sophisticated high-cost technology like GPS are freely available and used by most in daily life. People have witnessed the expansion of the health system and now benefit from advanced techniques like MRI and CT scan. Though much remains to be done with regard to access, we have managed to bring life-saving health care to those who were previously without hope. People born pre-80s find that several diseases that weren't curable before are now dealt with modern treatment methods. Even people who never imagined that they had a use for technology (such as the illiterate, semiliterate, and the aged) now carry smartphones as an essential personal good.

Politicians and bureaucrats who were previously skeptical of or disinterested in the transformative possibility of science and technology now see the evidence before their eyes. With our people becoming increasingly technology-literate and technology-aware, they can finally make a strong and even self-evident case for renewed focus and investment in scientific research.[20] Arguments that the disadvantaged would be better served with investments in social subsidies no longer carry the same weight when we consider how the poor have benefited economically from the ubiquity of technology. We need to balance short-term measures like subsidies, also a necessity for populist governments, with long-term initiatives like scientific research. Our leaders can finally speak persuasively to motivate people for research, to reform and develop the research ecosystem, and to allocate funding sufficient for a major national push. If Mr Modi can make an initiative around "Startup India," then he can make an initiative around "Scientific India." Although he might need a better slogan!

Another important reason to do it now is because we can. India is the world's third largest economy and the fastest growing economy. Economies as large as ours can provide the critical mass of people, the necessary investment, and the audience of users that are requisite for a great innovation ecosystem. With our size of economy, government and private players could fund research initiatives without abandoning other duties. There is enough for risk capital. For example, India could invest in the creation of a few world-class institutions of learning while maintaining funding for our existing universities and institutes. We are not only well suited to do research, we are well suited to enjoy the fruits of research. We have a large workforce waiting to be turned into researchers, innovators, and entrepreneurs. Our size allows labor specialization and development of multitude of skills possible. We have a large local market and testing ground for experimentation, pilots, feedback, and, eventually, large economic returns.

And finally, this is an opportune time to pursue scientific rejuvenation because we now have the chance to catch up with the West. Communication technology allows easy access to information and knowledge.

When my father did research in the 1980s at Delhi University after completing his PhD in the United States, he and his colleagues waited for months to see new research results. They had to wait until the journals arrived by boat! By contrast, when I was an undergraduate at NSIT, I had access to a plethora of papers on the web. I could contact authors directly by e-mail. People far away provided me with clarifications, feedback on my ideas, and provided internship opportunities. Moreover, NSIT had an online subscription to IEEE Explore, which offers quick access to papers. Today, we also have massive open online courses (MOOCs)! Most universities in the United States, including the best, have put their courses online, many free of charge. Our young talented minds do not even have to leave India to learn from the best professors in the field. They get their assignments graded, doubts cleared, and even earn a certificate remotely.

And let us not forget about the physical resources needed for research. These have also become more accessible and cheap, through technological advances and economies of scale. Processing power is now readily available on personal machines, smartphones, and even the cloud. Open source code for, say, operating systems and machine-learning algorithms are freely available for use in one's research. We also have open source repositories of biological parts (biobricks.org), circuits, and 3-D prints. Three-dimensional printing has made component manufacturing much easier, accessible, and cheap. There are several such examples that have made research less resource intensive and easier—making it easier for us to stand on the shoulders of others to see further ahead. All that is needed now is passion, creativity, and hard work to conceive new ideas and see it through to completion. A genius in high school could build her own 3-D printer to print brail: an MIT applicant from high school whom I interviewed has actually built one!

Now is the time for reinvigorating our science and technology ecosystem—the people are united, the economy is strong, and we have the tools at our fingertips. In this democratization of opportunity, we cannot rest. Ease of access means opportunity, but it also means that our competition has become much larger!

Summing It Up

अष्टादश पुराणेषु व्यासस्य *वचन* द्वयं || परोपकाराय *पुण्याय* पापाय परपीडनं ||

Vyasa summarizes all the 18 Puranas very simply: Charity toward others accrues merit, and causing suffering to others constitutes sin. My summary of the six reasons why research is a national priority and a necessity for India is only slightly more complex.

Scientific research will have great results for India. We can no longer do without high-quality research; the potential for doing impactful research is greater today than ever before; and this unique opportunity to catch up will not last if we do not take advantage. Following are the fruits of scientific research for us and our strength that shall enable us.

Fruits:

1. Scientific research can help us build innovative companies that create irreplaceable products and services with great

economic value. As owners of research, we can accrue the economic benefits of technological disruptions.

2. India-based research will produce public goods at lower prices. This is not just about money and profits. We can offer the world cheaper and more accessible health care, education, sanitation, and dignified living.

3. India-based research will position us to survive and excel in a winner-takes-all environment, where there is no room for second best. A winner-takes-all market implies opportunity, not limitation. We stand to gain through the development of specialized products of narrow application and targeted appeal that we can market to the world thanks to communication technology.

4. The innovation ecosystem will train a better breed of scientifically minded people, knowledgeable of contemporary developments, discerning, and capable of making wise decision. The proper ecosystem can create public intellectuals and a culture of intellect to competently guide the nation into the exciting future, as captains of the ship rather than passengers.

Strengths:

1. Our greatest strength and most wasted asset is our scientifically minded people. We are well advised to lead them to research careers with the right environment and resources. We owe it to ourselves for our future and to them for their ingenuity.

2. Our large and expanding economy provides the critical mass of people, investment, and audience for a great innovation ecosystem. We are well suited for research, for building innovative companies, and for enjoying the economic rewards. Globally, countries with large economic growth have invested in research and innovation.

3. We can produce impactful research at lower cost, giving us a competitive edge against the West. Our space program

operates on less than 1/10th the budget of other countries' space programs, and we produce the highest number of journal papers per GDP per capita.

4. The materials and hardware for research have never been more accessible and cheap. Thanks to the Internet, communication technologies, economies of scale, and the open source movement, we have everything we need to catch up with the West.

5. The people are ready. They have each personally witnessed and held in their hands the great benefits that technology brings to their lives. We can finally make a strong and convincing case for investing in and reforming our scientific research ecosystem.

3

Research in India: The Past, Present, and Future

If you cannot measure it, you cannot improve it.

—Lord Kelvin

India has achieved something that heretofore has only been done by the United States, the Soviet Union, and the European Union (EU): send a mission to Mars. However, we hear Narayan Murthy (founder, Infosys) say that India has had no "earth-shaking invention" or one that has "become a household name in the globe" in the last 60 years. Now and then, someone throws in a spanner that India could transplant heads thousands of years ago. A Nobel Laureate once called the Indian Science Congress a circus!

It is unfortunate to hear India's contributions to science disparaged in this way. Science is all about objectivity—science acknowledges the subjective and attempts to find an objective basis for the subjective. Is it proper for us to describe science itself in such a nonobjective frivolous manner? No, it is not, but we must admit that such talk is symptomatic of the dim state of scientific research in our country today.

We need to find the truth among all this noise. Objectively, where does India stand in research output in science and technology today? Are we doing better or worse than our past performance? Do we do great research or just good research? Are we better in some areas than others? How do we even begin to address these questions and plan for the future?

If we understand how research output is measured and how we identify what qualifies as great research, then we will be off to a good start. Additionally, we need to understand the meaning of the terms, "papers" and "citations," which we will introduce in this chapter. We will take a data-based approach to this. Simply put, we must "count:" count the number of papers India produces across various fields, how many papers are actually influential, and how we compare with other countries. This will provide us with a clear picture of where we are, how far we have to go, and how we can get there.

I first learned about the power of data through an organization called Coalition to Uproot Ragging from Education (CURE). I cofounded CURE in 2001, just after my freshman year in college. Those were interesting times: our primary challenge was that most people did not consider ragging (hazing) to be a problem at all! It was equated with the easygoing teasing and fun that is common between cousins. We wrote many articles about how ragging is against human rights and dignity; that it is nonconsensual and leads to injury and depression; and that in the long term, it can make people apathetic and inhumane. Our appeals barely caused a ripple.

We revisited our work in 2008, following my exposure in economics and machine learning at MIT. This time, rather than rely on shaming and moral exhortations, we took a data approach. We counted. We counted the total number of ragging incidents reported in the English media over the previous year; how many of these incidents comprised sexual abuse; how many led to injury or hospitalization; and how many resulted in death. We found 89 cases, of which 21 percent comprised of sexual abuse and 43 percent resulted in injury. In these 89 incidents of ragging, 11 people died. Interestingly, a substantial number of ragging incidents (31 percent) were reported from engineering colleges.[1] We constructed a simple table (Table 3.1) to present to the Supreme Court-appointed Raghavan Committee on ragging. In this chapter, we will see many such tables in the context of research.

Table 3.1: A Data-based Approach to Understand Ragging

Statistics	Academic Year 2007–2008 (Post-Supreme Court Order)	Annual Average over Last 5 Years (Academic Years 2003–2008)	Remark
Number of cases	89	46	Doubled
Number of deaths	11	5.6 (average)	Doubled (SIGNIFICANT)
Form of ragging	21% Sexual 43% Physical	25% Sexual 44% Physical	No change
Place of ragging	Engineering: 31% Medical: 17% Others: 52%	Engineering: 32% Medical: 17% Others: 51%	No change
Police intervention	50% cases	54% cases	No change

We circulated these statistics widely among the media; almost every national newspaper published them. This single table and our simple tabulations had a stronger impact than all our social science articles spanning 10 years. We like to think that we were responsible for the complete change in discourse and public opinion on ragging: thereafter, no one could claim that these acts were either isolated or benign.

In this case, we used data as an instrument to illustrate what we ourselves already knew: that ragging is widespread, harmful, and must be stopped. The dissemination of this data forced the change in social perceptions that we had hoped to see. However, data does more than quantify the obvious. Data provides insight, which inspires the right kind of intervention. Going back to the example of CURE, we determined that the highest percentage of all ragging incidents (31 percent) were happening in the engineering colleges. This suggested that antiragging campaigns would have optimal effect if concentrated on engineering colleges. Engineering colleges comprise less than 15 percent of colleges in India. One could focus effort on them for large impact than uniformly targeting all types of colleges.

We have still a third goal for quantification: improvement through "continued and swift measurement." "If you cannot measure it, you cannot improve it," was supposedly first said by Lord Kelvin and then independently rediscovered by me and several others over the course of the next hundred years. By measuring where we are every year, we can determine what has improved, what has not, and where further intervention would be most and least effective. We can determine what goals we need to set for ourselves. For example, in our report on the engineering profession for the most recent National Employability Report, we documented the unfortunate fact that there has been no significant improvement in the job skills of engineers. We don't need expert testimony to know that this is true: simple counting tells us so. Similarly, in our slick table on ragging we report on whether ragging incidents have declined since the Supreme Court judgment. Thus, these counts help measure impact, act as a "watchdog" and also create competition among institutions to improve. Quantification, therefore, is not an end, nor is it merely numbers exiled to an appendix; rather it is a starting point to better comprehend an issue, identify solutions, set goals, inform methods of approach, and measure improvement over time.

These are among the many benefits of data:

1. Finding evidence for a hypothesis.
2. Acquiring insight to inform interventions.
3. Measuring the effectiveness of solutions over time to determine impact, adjust goals, and create incentives to improve.
4. Influencing public opinion.

In fact, the data-based approach is becoming a standard to objectively report on issues of public interest. In India, we have data.gov from the Government of India for information statistics. The US government maintains census.gov and bls.gov; the EU compiles data at ec.europa.eu/eurostat; and even China is putting out national statistics at stats.gov.cn. Additionally, private companies and organizations maintain statistical data for various parameters and purposes:

indiastat.com, infoplease.com, socialcops.com, and statista.com are some examples.

Before we begin counting India's research output, we must admit that all data and metrics are incomplete in some way. They count that which can be observed objectively but might obscure other influences. We must approach numbers with discretion and caution, and within a set of clearly stated assumptions. Otherwise, we risk spawning rumors such as the recent news article on the resurgence of Bihar[2] (based on Aspiring Minds' report that engineers from Bihar had some good job skills!)

What Are Papers?

The research world requires some way to qualify new research—something that is useful, innovative, and has not been done before. Who decides whether a researcher's work is useful and unique? The research community establishes journals and conferences. Journals are similar to magazines, published once per month, or perhaps quarterly or biannually, and they contain research articles (papers) rather than general interest entertainment or lifestyle articles. Conferences are events—usually annual events—where researchers come together to present their research work, in the form of papers, to their peers. Every journal and conference has an editor or chairperson leading a committee of respected folks from the research community. Each paper submitted to a conference or journal is read by 1–5 reviewers; the editor or chairperson decides whether to accept the paper based on the recommendations of the reviewers (with or without revisions). Based on their expert knowledge, these people evaluate the paper to decide its merit, and whether it is worth disseminating through publication. This process is called "peer reviewing." This is an important step in the research process; we will come back to this step many times in this book.

I published my first research paper in 2003, in my third year at Netaji Subhas Institute of Technology. The topic was "automatically synthesizing sinusoidal oscillators using genetic algorithms." The

NASA/DoD Conference on Evolvable Hardware[3] invited me—an unknown student from nowhere—to Chicago to present my work in the final time slot of their 3-day conference. From sheer objective optimism, I say that they were saving the best for last! John Koza, arguably the inventor of genetic programming, approached me and complimented my presentation, remarking that he and his colleagues always found it hard to synthesize oscillators automatically. Then he gifted me his book. This is what conferences are for: to meet new people, learn about their work, build collaborative partnerships and friendships, and collect free books (how I gathered the funds to make it to this conference is a story in itself—very relevant to our discussion here, but which can wait until a later chapter).

Turning away from my insignificant contribution, let me describe a few examples of papers that are truly groundbreaking and the prominent forums in which they first appeared. The algorithm to find whether a number is prime in polynomial time, a long-standing mathematical problem, was discovered by Manindra Agrawal and his students at IIT Kanpur. This super-influential paper appeared in 2004 in the Annals of Mathematics. Watson and Crick first told us about the helical structure of DNA. Their research appeared in 1953 in the journal *Nature*, one of the topmost journals for breakthrough scientific research. Going back further, Jagadish Chandra Bose presented the construction of radio wave detector at the Royal Society in 1899.[4] His work was arguably used by Marconi to send radio waves over continents, from Cornwall, England to Newfoundland.

Citations Measure Influence

All the papers mentioned above are "highly cited." What does this mean? It is safe to say that all new research draws on previous research. Every scientist and researcher is indebted to those who came before. Therefore, when the researcher presents his or her work in a paper, he or she includes references, or "citations," to the earlier work that inspired or informed the new work. For example, my paper on oscillators referenced, or "cited," 15 earlier papers that had used algorithms to automatically synthesize circuits. By referencing them, I was telling my audience that this is an important field of work—that I am one of the latest in a chain of scholars who have paid attention to this question, that I have studied and synthesized their earlier work and have discovered something new to say. The new point in my work was that no one had previously discovered how to automatically design oscillators. And if my discovery is not merely interesting but also useful—as will be determined by time alone—then some future researcher will cite my paper in turn. Every paper cites other papers and could be cited by others. The research community meticulously documents and tracks citations. I cited 15 papers in my paper; since 2003, my paper has been cited by 21 others. As a point of reference, Watson and Crick's paper on DNA structure has been cited 11,581 times.

In this way, citations have become the accepted means of measuring the quality and influence of a paper. Even Google uses citations in a similar manner as a key parameter in its page-ranking algorithm—the number of pages that refer to a particular page. Citations are a direct measure of influence: more citations mean more people have read your paper and have found it valuable enough to cite. On the other hand, some people question the reliability of citations as indicators of value, and we will discuss this controversy twice, later in the chapter. For now, however, we can accept that citations are a good way to judge the influence and quality of a paper. One could employ a committee of experts to judge the worth of a paper, but citations is a more objective data-based way to decide.

People have invented various mathematical normalizations to get citations to reflect influence better. For instance, to compare papers across fields, we need to normalize the citation count by the total number of papers in the field. Otherwise, papers in a more popular field such as computer science will all seem better than papers from a less popular one.

Good Journals Versus Not So Good Ones

Not surprisingly, good papers go into good journals and good conferences. Just as there are good politicians and bad politicians, so are there supergood academic forums that accept the best of innovation, and poor outlets that will accept just about anything. In recent times, thanks to the Internet, there has been a deluge of low-quality and even fake journals and conferences. In 2011, a librarian named Jeffrey Beall started compiling a list of fake conferences and journals. He began with 18. Today his list contains 693 journals that he has identified as either low quality or fake. Fake conferences would typically modify the name of a top conference slightly to provide a perception of high quality.

For us, it is important to discern which journals and conferences are good and which are truly great—not least because we wish to know how often researchers from India rise to this or that category. As with research papers, the metric for rating journals and conferences is the number of citations received by the papers they produce. A great journal knows its track record and will only accept highly influential papers that are sure to be cited at a high rate. In turn, researchers know that they can look to such outlets to find high-quality papers and thereby stay abreast of the latest exciting news in their fields. The editor or chairperson decides the quality of papers that they are willing to accept. They also decide the acceptance ratio: papers accepted to papers submitted. Over time, most journals and conferences have settled into a hierarchy, wherein each one is aware of the quality of papers that they can reasonably expect to see. Mature researchers therefore have a good sense of which journal or conference might accept their papers; so they submit there.

Journals are ranked according to various mathematical citation-based measures with names such as impact factor (IF), SCImago Journal & Country Rank (SJR), and SCImago Journal & Country Rank 2 (SJR2). Each provides similar ratings and rankings, with a few differences and nuances. For example, two prominent journals *Nature* and *Science* have IFs of 41.45 and 33.61, respectively. *Physical*

Review Letters and *International Materials Review* (applied materials) are very good journals with IFs of 7.51 and 8.50. By contrast, the editors at the *Journal of Natural Products* and *Life Science Journal* are undoubtedly respectable people, but with IF factors of 0.165, their journals are of no particular consequence.

In summary, the research done in a country is largely measured by the number of:

1. Research papers published
2. Top journals and conferences they were published in
3. Citations contained
4. Citations received
5. Highly cited papers

These are the metrics we shall consider to determine where India stands today in the area of influential scientific research. As I said earlier, we will just count and compare.

Citations: Good or Bad?

Unfortunately, we know of no perfect metric of measurement. Yet we need measures to guide us and to help us correct things. Do you use a blood pressure-measuring device or a blood sugar-testing device? To some degree, we entrust our lives to these machines, but they all contain errors. Our citation measures also contain imperfections. Let us examine some of these issues and see if we can find a work-around.

A lot of Indian academics fume, if you mention that India doesn't produce highly cited papers. With such a start, you might either begin the relationship on an awkward footing, or you might abruptly end it. However, it is easy to understand the cause of their reaction: citations are a direct measure of influence, but not necessarily a direct measure of quality. A paper from an IIT might be of similar quality to a paper from MIT, but by virtue of its pedigree, the MIT paper will attract many more citations, and will therefore be considered more influential. Researchers will assign it more credibility just because the authors enjoy the MIT brand. Furthermore, MIT and its professors put great effort and resources into marketing their work. IITs cannot compete with such firepower, but that doesn't mean that papers from IITs are any less innovative or academically sound.

For this reason, we cannot consider the citation system to measure quality entirely accurately. Nevertheless, we should not discard the metric. Rather we should look for a way to correct it. We can do so by using secondary data. We can count the number of Indian papers that are published in the top journals and conferences. This will allow us to see if India does better in this metric rather than in the citation-count metric. If there is deviation, we could potentially correct it—for example, by increasing Indian citations by 40 percent to better reflect quality rather than influence. Top conferences and journals accept papers by quality, and they perform double-blind reviews (i.e., the reviewer does not know the name or nationality of the author, and vice versa). This measurement provides a reasonable—if not perfect—correction to the shortcomings of the citation measurement. We will revisit the citations issue at the end of this chapter, when I come full circle to argue why we must care for it!

But before moving on, we should point out that the inaccuracies in the citations system diminish for papers that can be classified as disruptive. Groundbreaking and even life-influencing papers such as "A Prime in P" or the paper that described the structure of DNA can come from anywhere, not just MIT. Regardless of their heritage, they will be recognized as disruptive and innovative, and will garner huge numbers of citations. We will examine later how India compares with other countries in regard to disruptive papers based on citations.

Our Research Influence: Counting Documents and Citations

Our paper and citation-counting process begins with Scopus, the largest abstract and citation database of research papers in the world.[5]

Their cataloguing process largely weeds out fake conferences and journals.[6] Researchers recognize Scopus as the most credible source for tracking publications and research trends. We begin by counting the number of papers (sometimes called "documents") and citations produced by India, and then we compare the numbers to the counts from other nations. We determine a paper's nationality by the location of the institute with which the author is affiliated (i.e., the paper's heritage does not depend on the country of origin of the researcher). A paper can be tagged to multiple countries, if the authors are collaborating from various institutes across borders.

Figure 3.1 lists the top 15 countries by number of documents from 2012 to 2016, a period of 5 years, and India ranks sixth. We produce around one-fourth the number of papers as China (rank 2) and one-fifth the number of the United States (rank 1). Looking at citations, India ranks 12th, just above South Korea. The highest number of citations belong to the United States, China, and the United Kingdom. Total citations from India are less than 10 percent

Figure 3.1: Documents and Citations of Various Countries for 2012–2016

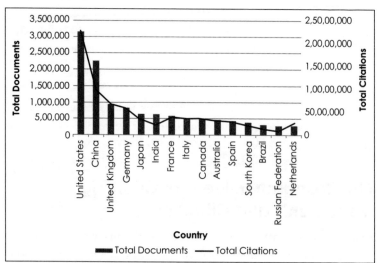

that of the United States, around one-fifth of China, and one-third of the United Kingdom.

India is not doing very well with citations. We are not among the top 10 countries on this metric. In comparison, we have the third largest university population in the world.[7] Our citation count with respect to the size of our educated population is very low. We have been able to get a lot more students to pursuing high education. But we have not focused on either the education quality or the research output.[8]

Our total research "influence" as measured by citations is low relative to the number of documents we produce. Our citations-per-document count is 3.39. In the United States, it is 7.15. China also performs a little better than India (4.29 citations per document) on this metric.

It might seem shocking that China ranks second in number of documents and number of citations. Many people still perceive China as a low-innovation, copycat-manufacturing economy. To the contrary, China has invested heavily in research and innovation over the last 20 years and appears poised to overtake the United States in research leadership. Thomas Barlow has written a book discussing whether the American Eagle or the Chinese Dragon will win the innovation race.[9] The Indian elephant does not deserve a mention, it seems.

What should be our place? We are the third largest and the fastest growing economy in the world, behind only the United States and China. In public discourse, we aspire to obtain "superpower" status alongside the United States and China. But as we discussed in Chapter 1, our aspirations will go nowhere without real leadership in scientific research along with an innovation ecosystem to realize the economic benefits of our research. In fact, as much as we are winning now, we will surely start losing if we do not ramp up our research efforts. India must become an innovation economy. To do so, we must become leaders in scientific research. Without these pursuits, we will begin our downward slide.

I firmly believe that we can join the elite. My conviction is based on the simple fact that if others can do it, so can we. We will see further, there are some pockets of greatness in India as well. India can rank among the best. For it to happen, we need to define research and innovation as national priorities. We recognize that our papers are of limited influence and have less influence than papers from the United States and China in particular. But how does influence correspond to quality? Are we producing good research? Let us consider this question now.

Research Quality: Standing Up to Be Counted!

Most researchers admit that writing and getting research papers published is not a mammoth task. As one learns about recent developments in a field, one starts seeing a lot of gaps, which can be addressed by some methodical thinking and creativity. Addressing such gaps with incremental solutions can add value: their experimental data and results can be of use to industry and might inform future research. However, some of us do not find these exciting. We prefer to spend our time investigating more challenging and interesting questions. We do not count ourselves on whether we are able to publish. We rather yearn the satisfaction to have solved a hard valuable problem, which makes a nonincremental addition to our current knowledge base.[10] Such results generally get published in top conferences and journals. Publishing at such avenues could serve as a better metric for quality than citations that measure influence.

To assess research quality, we will count the number of papers in top conferences and journals from authors in different countries. We consider machine learning, which is probably one of the hottest fields in both theoretical and applied computer science, a field that in my opinion holds much potential for India.[11] Then we pick mechanical engineering and pharmacology, two foundational and traditionally studied areas. We further include nuclear science, which is relevant both for our energy requirements and defense. Finally, we include biotechnology and neuroscience—revolutionary fields that are exploring foundational questions about the human species and helping to solve some of mankind's most pressing health issues. I first learned about neuroscience from a lecture by Professor Colin Blakemore (associated with Universities of Oxford and Warwick) organized by National Brain Research Centre (NBRC). It changed my way of thinking forever. One of topics included was "Phantom Limbs" (see "Phantom Limbs: Limbs That Are There but Not There").

Phantom Limbs: Limbs That Are There but Not There

I first heard about phantom limbs from Professor Colin Blakemore in 2004. Phantom limbs is a phenomena where a person with an amputated limb continues to feel pain as if the limb were still present. This phenomena is reported by 90 percent of amputees. The limb does not exist, but the mind of the person continues to perceive it: hence the "phantom."

In some pioneering work in the 1990's, V. S. Ramachandran, a professor at the University of California, San Diego, suggested a neurological basis to the phenomena. He hypothesized that the neurons that detect sensory input from limb were not receiving any signals from the missing limb. This led to the sensation of pain. This phenomena also made the neurons behave in strange ways: for example, touching the cheek of the person might give him the sensation that you touched his amputated limb instead!

More interestingly, Ramachandran developed a way to cure the pain in the missing limb. He put a mirror against the existing limb, such that it created the illusion of the missing limb still there. The patient looked at the illusionary limb and imagined flexing it, unclenching it, and massaging it, leading to a reduction in pain! In this way, the brain received the visual feedback of say the unclenching, which it could not get in the typical way, from the muscles of the limb. Ramachandran theorized that this was responsible for the reduction of pain.

Since then, there have been several theories concerning phantom limbs and the reduction of pain. But Ramachandran's work in this area proved disruptive. His papers on the subject—published in 1995, 1996, and 1998—have received 635, 932 and 1,045 citations, respectively. Ramachandran has continued to study how the brain works by studying other unusual phenomena experienced by people. He has also written three general interest books on the brain.

The brain plays tricks on us. Sometimes what we perceive is not really the "truth." I first experienced this breathtaking revelation in 2004. At one end, Eastern philosophies such as Vedanta have talked about this for 2,000 years. On the other, these ideas constitute new explanations that challenge the very basis of religious experience, through meditation, visions, and more. Our brain plays tricks on us. What is illusion and what is reality might be anyone's guess!

Table 3.2: India's High-quality Papers and Citations as a Percentage of Other Countries' Papers and Citations

Country	Papers in Top Conferences	India's Citations as a Percentage of Given Country's Citations	India's Papers in Top Conferences as a Percentage of Those of Given Country
USA	11,364	9.42	9.46
China	3,934	21.89	27.33
UK	3,039	31.48	35.37
Germany	2,853	36.09	37.68
Japan	2,150	66.47	50.00
Australia	1,149	67.65	93.56
South Korea	1,124	101.6	95.64
India	1,075	100	100.00

We have spent a century understanding the world outside (physics). In this century, we will understand the world inside (neuroscience and biotechnology). In each of these fields, we identified a few top journals and conferences through the Microsoft Academic Search ratings. To get a sample, we counted the number of papers in 7 countries out of the 12 listed earlier for the period 2012–2016.[12] We provide the aggregate numbers and ratios in Table 3.2 and Figure 3.2.

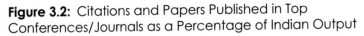

Figure 3.2: Citations and Papers Published in Top Conferences/Journals as a Percentage of Indian Output

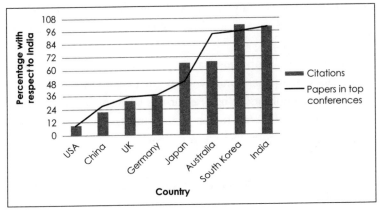

Looking at the total number of papers in top forums, India produces less than a tenth of the United States' output and around one-fourth that of China. Our counts are comparable to South Korea and Australia, but comprise less than half each of the United Kingdom, Germany, and Japan. By this measure, we are low on quality as well as influence. Can we apply a correction to the citation's index to better judge quality? We compare the citation index and the top conference papers index in the table. We find there is not much correction required in the case of the United States and China. The only outlier is Australia, compared to which India produces a similar number of papers in top conferences, but receives far fewer citations. So at least at the aggregate level, the argument that we publish good stuff but miss out on citations does not hold water.

Now let us compare different fields.[13] Compared to the United States and China, we do our best work in nuclear science and our worst in biotechnology and neuroscience. Our output in biotechnology and neuroscience is around 2 percent that of the United States. We produce 5 percent as many papers as China in neuroscience. Recall that these are the emerging fields of the twenty-first century!

Something must be well in our nuclear physics programs, as we manage to produce 36.4 percent as many papers as the United States and 54.4 percent as many as China. But in machine learning we are matching only 3.8 percent and 12.3 percent of the United States and China, respectively.[14]

It should now be obvious that we have a lot of catching up to do. Our citation and paper counts indicate that we are lacking, and our low count of publications in elite forums adds support to this conclusion. So next time an Indian person fumes when you mention citations, you can point them to this book!

Disruptive Research: Highly Cited Papers

Disruptive innovation, a term of art coined by Clayton Christensen, a professor at Harvard Business School, describes a process by which a product or service takes root initially in simple applications at the bottom of a market and then relentlessly moves up market, eventually displacing established competitors.

When we speak of "disruptive" papers, we are not merely speaking of those that are published in top conferences and journals. Disruptive papers are seminal. They solve long-standing problems, create new fields, spur groundbreaking innovation, and have the potential to upend industry and economics. In the history of science, the work of Newton and Einstein was disruptive, as was Charles Darwin's theory of natural evolution. In the history of technology, the steam engine that mechanized human effort was disruptive. The invention of vaccination by Edward Jenner marked a major breakthrough in medicine, while the development of anesthetic agents and chloroform in the 1800s revolutionized surgical practices. Today, the field of AI that is mechanizing human intelligence promises to be at least as disruptive. Sometimes disruptive papers aren't recognized in their times. (See "The Arrogance of Science.")

These examples are well known. Other examples, though not so commonly famous, were no less portentous. For example, Karmarkar's

The Arrogance of Science

We should mention an interesting point about disruptive research. Sometimes a paper is so far ahead of its time and so controversial that it does not immediately gain the recognition it deserves. Let us always be aware of the human element in the judgment of a paper's worth. For example, Bose's work on millimeter waves did not gauge much interest at the time. But now, hundred years later, this work is widely recognized for its profundity.

Another example is Jean-Baptiste Joseph Fourier, the person who first gave us the Fourier series. When he first submitted his work, his fellow mathematicians were not impressed. He presented his paper at the French Academy of Sciences in 1807 but no journal would publish it. Fifteen years later, in 1822, he compiled his work in the book *Théorie analytique de la chaleur* (The Analytic Theory of Heat). Still, his writings remained obscure until after his death. Today his book has 1,568 citations! The Fourier series and its derivatives form the basis of electronics and signal processing that we know today. Undoubtedly, there exist similarly disruptive papers today that will not be recognized until the scientific establishment catches up to its pioneers. Science sometimes suffers from its own arrogance. It is rather the arrogance of the science community!

algorithm to solve linear programs in nonpolynomial time in the 1980s held great consequence for the field of mathematics. Let me describe the problem that he was addressing and the efficiencies that resulted from his solution.

Let us take a simple example from the web:[15]

You need to buy some filing cabinets. You know that Cabinet X costs $10 per unit, requires six square feet of floor space, and holds eight cubic feet of files. Cabinet Y costs $20 per unit, requires eight square feet of floor space, and holds twelve cubic feet of files. You have been given $140 for this purchase, though you don't have to spend that much. The office has room for no more than 72 square feet for cabinets. How many of which model should you buy, in order to maximize storage volume?

In this question, one needs to maximize the total storage volume within the delineated space, while staying within the allotted budge. Such problems arise every day in business administration—whether designing a floor plan for a production unit or managing inventory. In the language of mathematics, we call it "optimizing a linear function subject to linear constraints." Before Karmarkar discovered his algorithm, one could not solve this problem for large number of variables and constraints—it just took too much time. But Karmarkar developed an algorithm to solve these problems that was exponentially efficient in time. His innovation resulted in many valuable contributions. For one, it solved a long-standing open problem in optimization that had thus far thwarted many fine attempts. Second, it created immediate economic value for industry and individual businesses and assisted several engineering fields that deal with linear programs. And third, it created "disruption" in the way people think about linear programming. His work led to the discovery of a class of methods called "interior point methods" which in turn inspired solutions to many optimization problems beyond just linear programs. To date, Karmarkar's paper has been cited 5,338 times.

India produced quite a few disruptive innovations before our independence. Jayant Narlikar (celebrated Indian astrophysicist) nicely describes some of them[16] (Table 3.3).

Of all these remarkable personages, I feel most close to Jagadish Bose. Recently Gaurav Gandhi and I found that Bose was the first to

Table 3.3: India's Disruptive Innovations Before Independence

Year	Scientist	Contribution
1895	Jagadish Chandra Bose	Wireless communication up to a mile; Semiconductor Crystal to detect radio waves
1920	Srinivasa Ramanujan	Number theory theorems
1920	Meghnad Saha	Ionization equation
1922	S. N. Bose	Particle statistics, behavior of photons
1928	C. V. Raman	Raman effect

observe and document the memristive properties (bipolar switching) of a device called coherer, which was used a hundred years ago to detect radio waves. Bipolar switching in thin insulating films has been a subject of much study since the 1960s. Memristor—considered the fourth passive element after resistors, capacitors, and inductors—was first conceived in 1971 and first observed in 2008, as reported in the journal *Nature*. We found that Bose had observed the phenomena as early as 1901[17] and had a physical explanation for it! He had built a mechanical X-Y tracer—similar to an oscilloscope—in order to experimentally plot variation in resistance. Oscilloscopes weren't available commercially at that time. Bose was undoubtedly way ahead of his time and the scientific community has since revisited his work several times to find new insights. Of all his work, my personal favorite is his demonstration that plants have life (i.e., they respond to stimuli). As with much groundbreaking research, it was widely ridiculed by the scientific community before it was finally accepted. Bose published several papers in *Nature*, and he achieved several disruptive innovations in his lifetime. His story holds lessons for us, and we will return to him again and again in our discussion.

It is worth asking whether India's contribution to research in the last 60 years been disruptive. Or a slightly different question, how disruptive has India's research been? Mr Murthy referred to this when he said there has been no research with a global impact from India in the last 50 years.[18] Dutifully taking up the challenge as a good public official, Dr C. N. R. Rao defended Indian research with "Let us not damn it all!"[19] Vijay Chandru, an MIT alumnus and an inventor of Simputer, then followed with an article on how to improve the impact of Indian research. He also said something that I disagree with: that any comparison to the MIT is unfair.[20] So then, what is the truth about disruptive research?

Let us look again at our favorite metric of citations—this time from another angle. Now we wish to examine how India ranks in the subcategory of highly cited papers. Papers that are disruptive are also very influential and will garner a very high number of citations. Think of it this way: If we wanted to compare the quality of students in two different sections, we would compare the grade average for each section. If the Class A grade is higher than the Class B grade, then we assume that Class A has better students. Seems obvious and uncontroversial. But if we look more closely, we might see that Class B has a Ramanujan, a Bose, and a Raman, while the rest of the section students perform quite badly. The grades of the lower performing students overwhelm the performance of the superstars, dragging the section grade below the grade of Class A. So even though Class B had some profound members, the overall average was bad. We would like to know if this is the case with India. Do Ramanujans, Boses, and Ramans lurk amidst the commoners?

To attempt an answer, we consider our seven chosen countries and the number of highly cited papers they produced from 2012–2016. The results are shown in Table 3.4.

In the last five years, India produced around 1,000 papers that are highly cited (>100 citations) while most other countries produced more than 3,000 such papers (barring South Korea and Japan). Even South Korea produced 1.5 times the output of India. The United States has 234 papers with 1,000+ citations, many more than any other country.

Table 3.4: Number of Highly Cited Papers by Country

Country/Number of Citations	>=100	>=250	>=500	>=750	>=1000
USA	15,000*	4,013	976	437	234
Australia	3,041	616	173	80	41
China	7,213	952	194	84	50
South Korea	1,505	330	100	56	34
India	989	169	39	21	17
Japan	2,307	429	110	63	39
Germany	6,573	1,069	272	130	69
UK	7,358	1,395	375	175	94

Note: *This is a lower bound on the number. The United States has so many such papers that it becomes hard to count them based on our methodology!

Table 3.5: India's Citations and Highly Cited Paper Count as a Percentage of Other Countries

Country	India's Citations as a Percentage of Given Country's Citations	India's Highly Cited Paper Count as a Percentage of Those of Given Country
USA	9.42	6.59
China	21.89	14.54
Germany	36.09	15.22
Japan	66.47	41.89
Australia	67.65	31.26
South Korea	101.6	60.99
UK	31.48	13.44

So how does our highly cited paper count compare to the citation count? In Table 3.5, Column 2 shows India's citations as a percentage of other listed countries. In Column 3, we see India's percent value of the highly cited paper count (papers with >100 citations).

By this measure, India performs worse at producing influential papers than in garnering citations. This is like saying we can produce fairly good athletes, but find it really hard to produce Olympians. In terms of citations per paper (average citations), we are similar to China (China's total citation count is four times higher than ours due to the sheer volume of papers produced in China). However, the situation looks different if we consider only highly cited papers. By this metric, China outpaces us by a factor of 6 and not 4! The Chinese are winning not just on quantity but on quality. China is indeed producing many more Olympians than we are. It's not just an allegory! They are proving themselves adept at creating pockets of excellence to rival the best in the world.

This comparison holds true across the list of countries. There has been impactful research from India and in the last 15 years, there have been 18 disruptive papers from India. Yes, we need to catch up, but these 18 papers show that we've got what it takes.

Who are our star researchers? I am proud to say that one is a personal friend, Dr Kalyanmoy Deb.[21] We have both published quite a few papers in the field broadly known as evolutionary algorithms. He invited me to chair the industrial session at Simulated Evolution and Learning (SEAL), a conference he organized in India in 2010. Like Dr Karmarkar, Dr Deb is also interested in problems involving maximizing or minimizing things. The paper that won him so many citations—and also the Shanti Swaroop Bhatnagar award—tackled a slightly different problem, that of multiobjective optimization. As the name implies, this involves optimizing more than one objective at a time. Let us go back to our earlier example where we wished to maximize cabinet space (storage volume) amidst constraints in budget and total floor space. Perhaps we wish to find the maximum number of cabinets we can buy at different budget amounts. For example, how much cabinet space could we get if we wanted to spend just $100 or $120? Based on how much maximum floor space I am allotted at each of these different dollar amounts, I will make my final decision. This can be useful in more than one way. For example, perhaps I can get sufficient space for $100, whereas each additional $10 does not get me much more. Then I may settle for the $100 allotment and postpone additional spending until there is a critical requirement.

In another example, perhaps I need to make cabinet-buying decisions for several different offices, each with a different budget. Traditionally, this would be a multiple linear optimization problem, wherein each phase is subject to a different budget constraint. However, with multiobjective optimization, minimizing cost and maximizing cabinet space are both objectives. Formulating the problem in this way gives multiple solutions akin to determining maximum cabinet space at differing budgetary limits.

Professor Deb discovered an efficient genetic algorithm to solve such multiobjective optimization problems. The paper was neat. It did not offer any new fundamental mathematical results, but provided a huge algorithmic improvement. This not only inspired a multitude of new algorithms in the field but was readily adopted by many companies in the design of various machines from furnaces to airplanes. Dr Deb has consulted for many companies such as General

Electric, General Motors, Tata Steel, and Honda. This is a very good example of a disruptive innovation that is inspirational for academics, but which also yields direct economic value. Unfortunately, we recently lost Dr Deb to the University of Michigan in the United States!

Our Research in Different Areas

So far we have been comparing how India's total research output compares with the output of other countries. It will be further useful to examine whether India performs better in certain areas than in others. Let us consider India's rank in the 27 different subject areas covered by Scopus. In Figure 3.3, we show India's rank by citations in these areas from 2012–2016.

By citation count, our top 5 subjects are pharmacology, energy, chemistry, chemical engineering, and material science. Our average rank in these areas ranges from 4.6–6.4 (which is much better than our

Figure 3.3: India's Document and Citation Rank by Discipline

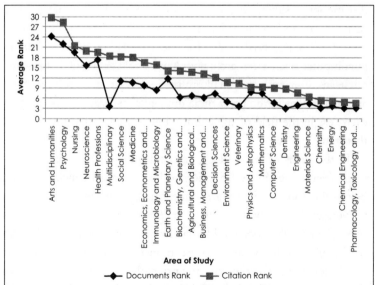

overall rank of 13). India's competence in chemistry, chemical engineering, and material science is well known. Researchers at the IITs and Indian Institute of Science (IISc) are key contributors in these fields, as are the scientists at CSIR, a network of government labs formerly headed by R. A. Mashelkar. Mashelkar is a highly respected (and highly cited) researcher who is probably best remembered for his efforts at reforming the CSIR system in the late 1990s–early 2000s. He is a leader in the campaign to promote quality research in India. Professor C. N. R. Rao, an advisor to the Government of India, is also a highly cited professor working in the fields of chemistry and materials. His affiliated institutions such as Jawaharlal Nehru Centre for Advanced Scientific Research (JNCASR), and others such as Indian Institutes of Science Education and Research (IISERs), are strong contributors to area research. Also, the government launched a Nano mission in 2007 to encourage research in nanotechnology, a subarea of material science.

Our performance in other areas is only average: computer science (9), genetics (14), and agriculture (13.6). Research in computer science and genetics is creating new knowledge and leading to significant disruptions, resulting in great economic value. Unfortunately, these benefits are mostly accruing elsewhere. Our thriving IT industry constitutes a potential outlet for the application of new findings in computer science research; if only we could generate that research ourselves. Our rank in agriculture seems out of step with our focus and investment in the field. India graduates its largest class of PhDs each year in the field of agriculture. I am curious what they do with their time.

In the fields of psychology, arts and humanities, social sciences, neuroscience, nursing, economics, health care, and medicine, we rank below the top 15. These are our worst performing areas. As the cofounder of a company that develops personnel skills assessments, I can well understand India's low rank in psychology. We have hardly any psychologists who are trained to understand numbers. Psychometry, which combines the principles of psychology and statistics, is a very well developed field in the West. It is mostly nonexistent in India. Neuroscience is a hot, new, cutting-edge field globally, with enormous promise. Our understanding of the brain has broken all bounds in just the last couple of years, commensurately informing us how much further we have to go and offering tantalizing suggestions of the discoveries that await. In pursuit of this, MIT has established two new research centers: The McGovern Institute for Brain Research (2000) and The Picower Institute for Learning and Memory (2002). These new entities are part of the Department of Brain and Cognitive Sciences, formerly known simply as the Department of Psychology. The Department of Psychology was founded in 1964. India has been sitting on the sidelines. Recently we have been trying to put our act together with thought leadership and funding from Kris Gopalakrishnan, one of the founders of Infosys. Mr Gopalakrishnan donated money to start the Centre for Brain Research at IISc in 2015. Prime Minister Modi presided over the inauguration. On the other hand, economics has recently become exciting again, thanks to new blood, new energy, and new ways of thinking. Developmental economics is going through a renaissance recently and is in particular relevant to India. Unfortunately, most breakthrough papers in the field come from authors in US and European universities.

Our performance in nursing, medicine, and health care is difficult to excuse. With our poor record on health care, these fields should be a particular priority. The Human Development Index, a composite indicator of human well-being, ranks India 131 in 188 countries. India needs progress in health care and medicine not just for global economic competitiveness but to solve our own problems as a matter of compassion and dignity. Some diseases common to India have unique characteristics, which are not addressed in other global studies. In addition, our scale of population, their diversity, and economic condition are very different from the West. So are many of our health science issues. And of course, by helping ourselves and solving our own health problems, we should not be surprised to find ourselves in a leading position to offer high-quality, low-cost health care to the world. Already, our low-cost generic medications have spread to many parts of the world. We can be proud of this, but we should be mindful that we are merely reengineering what other people have already developed. We need to create new knowledge—methods and medicine—that works better than those that already exist. Doubters should read the example of Abhay and Rani Bang (see "The Story of Rani and Abhay Bang").

The Story of Rani and Abhay Bang

I first got to know Abhay and Rani, when one of my colleagues at Aspiring Minds sent me an article written by Abhay on Mahatma Gandhi.[22] My colleague wished to point out the similarities between Abhay's article and an article that I had written on how I perceive Gandhi.[23]

Abhay's father was a Gandhian. He wished to go to the United States to study economics, but on the Mahatma's suggestion he took on learning economics by spending time in Indian villages. Ironically, and in contrast to the Mahatma's advice, Abhay went to Johns Hopkins University, where both he and his wife received their Masters in Public Health.

If at all, Johns Hopkins proved to be a temporary deviation from the Mahatma's way. The couple returned to spend time in Gadchiroli, a district in Maharashtra, working on community health issues. They basically converted Gadchiroli into a scientific laboratory where they ran experiments in public health, while serving the people.

The couple was pained by the high infant mortality rate among the residents. They studied the various causes for the mortality rate and came up with a "Home-based Neonatal Care" model. In this model, trained village health workers provided neonatal care in their home visits. The Bang's methods were completely unconventional. At this time, people believed that proper neonatal care (of the lifesaving kind) could be done only in hospitals. The couple's approach brought down the neonatal mortality rate by 62.2 percent, the infant mortality rate by 45.7 percent, and the perinatal mortality rate by 71 percent.

They published a paper on their methods and results in *Lancet*, one of the oldest and best known journals of general medicine.

To date, this paper has received 783 citations. Their paper on pneumonia treatment has garnered 226 citations. They halved the mortality rate from pneumonia in children under 5 (8.1 compared to 17.5 in areas where intervention was not performed). Their methods have been accepted and implemented by the Indian government on a large scale and also by developing countries around the world. There work is surely disruptive and an inspiration to many.

There are lessons to learn from Abhay. For example, one should not follow only the established methods of success, but try new ones most suited to the problem at hand. At the same time, one should produce irrefutable and unchallengeable results. If one wishes to produce disruptive new knowledge, one must have the courage, persistence, and readiness for the long haul. Rani and Abhay's neonatal experiment took them 5 years: 1993–1998. And they knew that there was no guarantee of success.

We need to find ways to produce more individuals such as Rani and Abhay Bang. The Mahatma is always there for inspiration. If the Mahatma had not stopped Abhay's father from going to the United States, Abhay might not have been influenced by the idea of service. And if Abhay and his wife had not decided to go to the United States, he probably would not have received the training required for the great work that he did. The path to innovation and greatness is nonlinear!

Research Productivity—When Will We Catch Up?

So far we have been discussing research productivity in India within narrow time frames. And we have found that within these time frames, India's record is not so enviable. However, we must consider something equally important: the growth of research productivity over time. The long view can give us broader insight into whether we are improving and whether our growth is relatively slow or quick. Growth is an important metric in any field. Economic growth in particular is among the most closely watched barometers of a nation's health.

Of late, India has been one of the fastest growing economies in the world. Does this good news hold true across other areas?

A faster growing entity will move ahead of a slower entity eventually, even if the faster entity begins further back. A speedier car eventually overtakes a slower car, even if the slower car had a big head start. Of course this may take time, but theoretically, and without limits in time and distance, the faster growing entity will always win. Furthermore, annual fixed percentage growth has a compounding effect. If an entity begins at 100 in year 0, and then commences to grow at a rate of 20 percent, it becomes 120 in year 1, 144 (120*1.20) in year 2, and 172.8 in year 3 (you add 24 instead of 20 in year 2, and 28.8 in year 3). Thus, if the entity can maintain a constant year-on-year growth rate, the entity will grow exponentially.

If the starting distance between a faster growing entity and a slower growing entity is very large, it may take years for the faster growing entity to catch up. Additionally, as entities grow larger, their growth rates decline due to saturation effects. America's economy is presently growing at 2.2 percent, while India's is growing at 7.3 percent. Nevertheless, we are not ahead of America. Thus, we should not deceive ourselves into thinking that we are doing better simply because we have a higher number. We need to consider growth in the context of absolute achievement.

Accordingly, let us consider the growth of research document output in India. Figure 3.4 shows document output for several countries from 2007–2016 (Figure 3.5 shows a magnified view and excludes the United States and China).

The United States outperforms all other countries in document output, showing steady growth from 2008–2012. We do see some plateauing, but it is still pretty amazing that the United States continues to grow even while maintaining such a "rate of growth." China has been growing stupendously and at a much faster rate than any other country. China almost doubled its document output in just 5 years, 2007–2012, though it shows plateauing in the last couple of years. It will be interesting to see if it can revive its growth once again. Nevertheless, we are indeed growing faster than any

Figure 3.4: Publications by Country, Annually

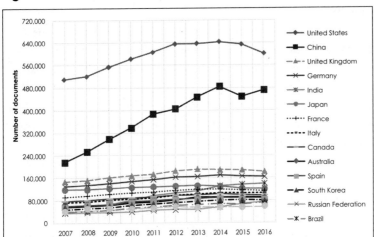

Figure 3.5: Publications by Country, Annually (Excluding the United States and China)

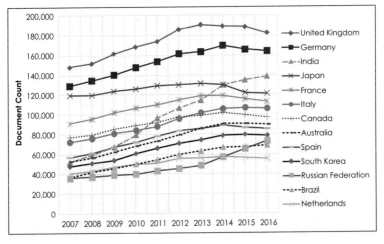

other country on the list, other than China. In the 5 years from 2007–2016, India added 85,000 new documents, while China added around 250,000. Three times more! In terms of percentage, both countries have comparable rates of growth, but it is still rather impressive that China can maintain such strong growth with a much larger base.

As before, however, we are more concerned with citations than documents. But we cannot track the growth rate of citations as easily as we can track the growth of document output. Papers that are recently published naturally have fewer citations on average than older papers; readers have had less time to locate and review them and to consider how to use the information to inform and support their own work. Because of this situation, the plot lines of citations will always decline year after year. Therefore, we must normalize the citation count for each country by the total citations for all documents in the given year. This gives us the citation factor for each country in proportion with the whole world, for the year.[24] Figure 3.6 shows the percentage citations of various countries from 2007–2016 (we have excluded the United States for clarity).

China's citation percent grew from 4.9 in 2007 to 11.8 in 2016. By comparison, India's citation percent grew from 1.6 to 2.4. China has more than doubled its citation percent and accounts more than 10 percent of world output. India has grown by 50 percent and still

Figure 3.6: Percent Citations by Countries, Annually

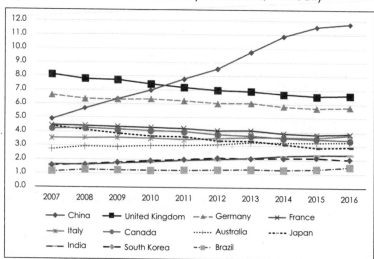

accounts for less than 3 percent of world output. If we look at citations in terms of US output, China has grown from 16.5 percent of US output in 2007 to 59.5 percent of US output in 2016. India, on the other hand, has grown from 5.3 percent of US output to 12.2 percent. We are the fastest growing country in terms of citations, but we are growing much slower than China and we are starting from a much smaller base.[25]

In terms of our growth rate in various fields,[26] our steepest growth rate has occurred in the fields of chemical engineering, chemistry, multidisciplinary research, material science, and engineering. Among these 5, multidisciplinary research is the clear winner; India's contribution began low but has been growing fast. However, in nursing and medicine, our contribution is low and apparently going nowhere. We show moderate growth in computer science, energy, and neuroscience. But we must admit that China is growing much faster in every field (except health). China seems to have a much higher focus on mathematics, energy, and decision science compared to India. And they apparently have a lower focus on social science and psychology (perhaps understandably) and also health, planetary science, and physics.

Beyond Papers and Citations

Number of papers and citations do not necessarily measure economic gains or how industry is using research for commercial products. They also do not necessarily measure the social impact. Should we not talk instead about systems such as the Internet, telescopes, and the Conseil Européen pour la Recherche Nucléaire (European Council for Nuclear Research; CERN) cyclotron? Why should we not count patents? Or the number of innovative products (iPhone-Calibre) that are developed in India?

Systems and experimental setups such as the CERN cyclotron, driverless cars, supercomputers, UNIX systems, language-compiling software, and circuit simulators are the fruits of great research. They have a systemic effect on the overall innovation environment in any country and the world: accelerating productivity of researchers, enabling collaboration, and leading to multidisciplinary research.

Do consider that papers and citation counts act as a proxy. Great systems generate a lot of citations. Note the 1,546 citations for Bach's UNIX book and the 320 citations for the book on SPICE, a circuit simulator developed at University of California, Berkeley. Some researchers generate great social or industrial impact, without garnering much citation. I hold such researchers in very high esteem and would not want to force the citation criteria on them. There are some such people in all institutions, fields, and nations. When we look at aggregate numbers for a nation, the attenuation in citation count because of such people may as well be consistent across the board (a systematic error).[27] Thus, the citations metric works well to measure performance at an aggregate level even if it is not the best metric for measuring the performance of certain individuals.

To get a direct peek, we will attempt to estimate India's contribution to systems in the next section, using some approximate measures. There is no listing of peer-reviewed projects or systems, unfortunately.

Let us now consider the economic value created by research. Research leads to the creation of new knowledge. This is a goal worthy in itself. However, the real tangible benefits come through the application of research that result in public goods or economic value. Great research is the first step in this process, which is why it is the primary focus of this book. Given that the public/economic impact of research in itself affects the research ecosystem and its development, I cover it in some detail as well.

Patents comprise one indicator of economic value associated with research. There are a lot of misconceptions about patents. Laymen might think that patents are testaments to great research and invention. People (like me!) flaunt the number of patents on their résumés: how many they have filed and how many they have been granted. Companies advertise the same. However, it is important to distinguish between a patent filed and a patent granted. A person can file a patent for any supposed invention, no matter how useless and flimsy. The filing process does not involve external review. If a person is willing to bear the cost, then she may inflate her resume as much as she likes. The actual granting of patents is another thing entirely. This process can take 2–5 years, is considerably more difficult, and involves an official review along parameters such as novelty and inventiveness. Even this long process does not bestow legitimacy. Unfortunately, patents are often granted for superfluous things, such as copies of old inventions with minor adjustments. Therefore, patents are an unreliable measure of research quality. Besides, theoretical results and discoveries do not qualify for patents.

Patents rather represent the importance one gives to research that creates economic value. The process can cost anywhere from $5,000–$10,000 for a single granted patent. Once granted, the patent makes money through licensing: any individual who wants to use the invention for commercial purposes needs the patent holder's permission and must pay the required fee. Given the price of the patent application, one would only commit to the investment if she believed that the invention would earn a good return and create high economic value either locally or globally. In this way, counting patents is a good way to measure the confidence of people within a country that their work has high business potential.

Let us now discuss India's contribution to systems and patents.

Systems and Patents

The first open-source system I used during my undergraduate studies was SPICE, a circuit simulation tool built at University of California, Berkeley. All circuit companies use circuit simulators on a large scale. Most of these simulators have its roots in SPICE. I used another one called topcap, built by Professor James Grimbleby, University of Reading, UK, for my paper to automatically invent oscillator circuits. Since then I have used tools such as R, a programming language for statistical computing, LIBSVM, a tool to build machine learning models, OpenCV, a computer vision library, and HTK, a speech recognition engine, in my work at MIT and Aspiring Minds. R was conceived at University of Auckland, New Zealand, LIBSVM in Taiwan, OpenCV by Russian contributors in Intel, and HTK in University of Cambridge in the United Kingdom. None of these were from India.[28]

To get a little more objective, we looked at some categories mentioned in "List of Free and Open-source Software Packages" on Wikipedia.[29] Among 5 packages listed for AI, 16 listed for Machine Learning, 6 in Encryption, 124 content management systems, and 78 IDEs, there is only 1 from India, in the content management category. Among the list of packages in the "Machine Learning" category, we find 9 are from the United States, 3 from New Zealand, and 1 each from France, Spain, Slovenia, and Germany with an interesting mix of university student/faculty and corporate contribution. There is none from India and China.

This data is representative of India's poor performance in building research-led systems. In fact, it is probably poorer than it is in papers. It is hard to build good software that can be used for a wide number of use cases, without breaking. You could appreciate this from the experience of the number of bugs you may encounter in any software, even in Windows and Android, built by respectable companies. It also needs a culture to share the confidence of sharing one's code and consider it as good research to build open-source tools for new algorithms. India needs to develop this big time.

Let us next consider patents. According to the 2016 report of "World Intellectual Property Indicators," India ranked 14 among patent-filing activity, while China ranked first and the United States second. China had more than a million applications filed, the United States 600,000, and India 45,000. We are a magnitude smaller than China and the United States in the number of applications we file. In terms of growth in 2010–2015, China has grown by 220 percent, India by 60 percent, and the United States by 20 percent.

Let us consider the Patent Cooperation Treaty (PCT) application, which helps initiate the process to get protection in multiple countries. One would file a PCT if they believe that they have a global innovation and want to protect their business interests in multiple locations. Here, around 30,000 PCT applications originate from

China, 60,000 from the United States, and a mere 1,500 from India. We are 20–40 times smaller than the United States and China.

Let us dig a little more. Among the top 100 patent applicants worldwide, no companies or institutions from India feature. There are 8 Chinese universities and 8 Chinese companies in the list. There are 8 entries from the United States. The top 5 PCT applicants are Huawei Technologies (China), Qualcomm (USA), ZTE (China), Samsung (Korea), and Mitsubishi Electric (Japan). There are no Indian companies or universities among the top 20.

Clearly, India doesn't do any better in patents than publications. This indicates that our inventors and businesses do not see innovation as a competitive advantage. We do not think it can create great economic value for us. On the other hand, both China and the United States see themselves as economies based on innovation.

Citations—A Necessary Good NOT Evil!

You might be feeling weary from the deluge of citation statistics in this chapter. While there may always be conflicting opinions on citations, they occupy a significant place in research measurement. I predict that this will not change anytime soon. Citations will remain a key metric for judging research productivity at an aggregate level. Even for us, they refuse to go away as a topic of discussion.

Sometimes papers with high citation counts are actually substandard. Some researchers band together in coteries: they produce substandard or even useless research and then cite each other's work. This artificially inflates their citation counts to make their papers look good, although the papers might be lacking in intellectual substance or are otherwise inconsequential. It is like an exclusive club of poor researchers! Some people do this because they have been trained poorly, or they are desperate to fulfil some institutional requirement for promotion or renewal. In other cases, the particular field or area might have been impactful and interesting in the past, but has since lost its shine. One such field is linear circuit theory, especially in the area of small circuits. This field was quite dynamic

25 years ago. However, the field was hollowed out by the advent of integrated circuits, transistors, and the need to analyze ubiquitous nonlinear circuits. Automation took over much of the work that had been done manually by researchers. I personally witnessed this phenomena through my work in the automatic synthesis of oscillators. Yet we still see several papers each year wherein each researcher presents a manually designed oscillator. My algorithm could design all these oscillators automatically. Moreover, the oscillator topologies on which these researchers labor have no relevance today to any practical circuit design. Indian institutions and scientists need to rescue themselves from such obsolescence.

Citations might be exaggerated for papers from the United States, even if not at an aggregate level. Citation exaggeration is probably more common at the individual level and in some fields more than others. In 2006, at the Genetic Programming Theory and Practice Workshop, I met a professor from a top US university who presented on a supposedly new method of fitness calculation for genetic algorithms. After the session ended, I remarked in the presence of a few others that similar work had already been done by researchers in Singapore, who termed it "surrogate-assisted evolutionary algorithms." He politely claimed to be unaware of the work and asked me about it. I had been holding a paper in my hand. I flipped to the back pages and said, "All these papers mentioned in the references section are the group in Singapore." He was clearly displeased. He had conveniently missed the work that an entire research community in his own field had done over years.

There are several reasons why this happens. People conveniently do not search the literature well enough. Possibly, for good reason, they don't want to know! I estimate that at least 10–15 percent of so-called "good" papers are just reiterations of papers that have already been published. Elitism is also a factor—researchers in the United States and Europe assume that nothing of importance is happening elsewhere and so they do not bother to look (other than to China). An outsider must really shake them hard to get them to take notice. And, of course, part of the problem is us. The world does not see India as a place for quality research or innovation. We are more commonly seen as a destination for cheap IT and business process outsourcing (but at least we are no longer the land of cows and elephants!) We suffer from Indian stereotypes, but it is up to us to break those barriers and change those perceptions.

And this leads me to my point. We cannot content ourselves by attacking citations and arguing that they are not a good measure of quality. Strange to say, we need to concentrate on more than just quality. We need our research to be influential and impactful as well; those things will not necessarily follow directly behind quality. We need to become good marketers and speak loudly enough so that our research gets noticed, and so people can no longer neglect to cite us without embarrassment. Research that is not diffused is like research that never happened. Research must become known and must cross-pollinate with other ideas for it to generate economic value, serve the public good, and perpetuate itself by generating funds for future research. Progress happens when one researcher becomes intrigued by the work of another and then uses it to inform and further his or her own investigations. This process is an integral part of the research ecosystem. We need to participate in it, and not watch from the sidelines. Research awareness and reputation leads to more and better opportunities for collaboration. When industry and government take notice, then there is the real chance for economic value and social benefit. Eventually, as India improves its reputation, it will become easier for us to get noticed and get citations with lower effort.

How do we promote ourselves? The best way is always face to face. We must become apostles of Indian research, travel widely to groups, workshops, and conferences, and proclaim the truth of things. After speaking at Microsoft Research Redmond, my team got our programming work noticed, and a short time later our citation count picked up. I co-organized workshops with some US authors at machine-learning conferences. These colleagues also took notice of our work. We also invite researchers to come to us to give talks at our headquarters, resulting always in a lively and productive sharing of ideas. We maintain a mailing list of all active researchers in the field, who we directly inform about our new results by e-mail every now and then. These efforts help us increase our sphere of influence and attract citations, collaborations, and also business. However we choose to do it, the simple fact is that we need to spread the word about India's research. Traveling, meeting, and presenting are ways to do it. This is another big problem with India that we will discuss; stay tuned.

4

Researchers: Attracting Our Best Minds

By far the most dependable indicator of university status is the faculty's degree of excellence that determines nearly everything else: a good faculty will attract good students, grants, alumni and public support, and national and international recognition. The most effective method to maintain or increase reputation is to improve faculty quality.

—Henry Rosovsky, ex-Dean, Faculty of Arts
and Sciences, Harvard University

In 2007, McKinsey conducted a study to compare the quality of primary and secondary education among various countries. They determined quality by measuring the students' performance on achievement tests in math, science, and reading. Finland topped the list. In Finland, the highest achieving college graduates choose to become teachers. As the most intelligent and wise, they are best able to lead the young students to their highest potential. Teacher quality was identified as a primary factor in student success. Therefore, McKinsey concluded, "The quality of an education system cannot exceed the quality of its teachers."

In a similar way, the quality of research output is constrained by the quality of researchers. In fact, we might argue that the relationship between the quality of manpower and that of output is strongest in the field of scientific research than in any other endeavor in contemporary human activity. The seat of innovation is the human mind: this is where the "work" that produces great innovation takes place. If the people who are tasked with performing research are not the best among us, no secondary inputs can compensate for it. Just as competitive sports test the limits of physical endurance, strength, and mental agility, scientific research tests the limits of cerebral acuity and acumen. Henry Rosovsky, the former dean of the Faculty of Arts and Sciences at Harvard, says in his book about hiring research faculty: "we ask the traditional question: who is the most qualified person in the world to fill a particular vacancy, and then we try to convince that scholar to join our ranks." We need talented researchers who possess passion, intelligence, tenacity, and have a history of high achievement in their field. Without these, we are unlikely to produce groundbreaking or even notable work.

When I come across shoddy research papers, I joke with my research team and repeat Mark Twain's wisdom: "Never try to teach a pig to sing. It wastes your time and annoys the pig—similarly, don't let dull people do research. It wastes everyone's time and money and depresses the person." Far be it for me to call anyone a pig! Nevertheless, we should allocate talent by which we can have the greatest impact on innovation and our economy, while achieving self-fulfillment and happiness along the way for the individual. Research needs to be the vocation of those who are most scientifically inclined, who have demonstrated excellence in their field, and are among the top achievers.

Having high achievers do research is not enough. Society must provide the required infrastructure and environment as a duty toward the high achievers if we hope to realize the benefits of truly innovative research and progress. The environment is just as critical as the researchers are. We need both.

In this chapter, we focus on the researchers. Who are these researchers? If I asked the average person to name the top industrialists in India, they could probably produce half a dozen: Birlas, Tatas, Ambanis, etc. If I asked for the top sportsmen, there would be a deluge of names: Tendulkar, Dhoni, Abhinav Bindra, and Anand. Most people could identify some famous entrepreneurs, such as Sachin and Binny, Kunal Bahl, Ritesh Agarwal, and Vijay Shekhar Sharma. But who knows the names of our best contemporary scientists and researchers? A. P. J. Abdul Kalam might come to mind, but probably for his much larger contributions than his scientific exploits.

Unfortunately, society does not celebrate our researchers. We are mostly unaware of who our researchers are, what they do, and who has made it big by helping create the things we take for granted in life. The fact that we have no role models to inspire young people for research careers indicates a problem with our national orientation. This is the subject of this chapter. Let us begin with knowing more about our research community, their numbers, what they do, and how well they perform.

Counting Researchers

Research is conducted in a variety of settings. Universities come first to mind, but government and private business also maintain laboratories for scientific pursuits. These labs chiefly sponsor PhD-level[1] scientists who work in teams to solve research problems. Indian Space Research Organisation (ISRO) and India's DRDO, plus the many CSIR Labs, are good examples of government labs. In the corporate sector, GE Research, IBM Research Labs, and TCS Innovation Labs are particular standouts. Then there are innovative companies such as Strand Life Sciences and my own Aspiring Minds that conduct research, publish papers, and file patents. Despite the diversity of the research community and the democratization of the tools necessary for performing research, most scientific inquiry still occurs in universities and other institutes of higher learning, led by faculty and supported by teams of students. Universities comprise the center of research training other than producing high-quality research output. Stanford, Caltech, and MIT, just to name a few, have been responsible for many of the greatest innovations of the

Table 4.1: Total Research Personnel in India

Sector	Research Personnel
A. State/central government Labs	87,905
B. Higher education Institutions	22,100
C. Industrial sector (public and private)	82,814
Total	192,819

Source: DST, 2010.[2]

twentieth century. Our own IISc and Tata Institute of Fundamental Research (TIFR) are hopeful peers.

Let us first consider the number of researchers we have and how these numbers compare to the United States and China. Table 4.1 lists the number of researchers in various settings in India in 2010.[2]

According to these statistics, India has approximately 190,000 research personnel, largely working in government and industry. Higher education accounts for only 11 percent, a small portion of the total.[3] How do these numbers compare to the United States and China (Table 4.2)?

India is far behind in quantity of researchers.[4] The United States and China have 1.4 million and 1.1 million, respectively, seven to eight times as many as India! If we consider the number of researchers per 1,000 educated people, China beats us by a factor of 3 while

Table 4.2: Number of Researchers, Educated Population, Workforce, and Ratios

Country	Total Researchers	Population with Tertiary Education	Ratio: Researchers/ 1,000 Educated People	Organized Workforce	Ratio: Researcher per 1,000 Workers
India	192,819	68 million	2.8	30 million	6.4
China	1,152,311	140 million	8.2	-	-
USA	1,412,639	124 million	11.3	100 million	14.1

Sources: Column 3 and 5 assembled using data from World Bank, CIA World Factbook, India and US Census, NSSO.

the United States outpaces us by 4. Normalizing these counts by organized workforce, we have less than half the number of research scientists as the United States. Not only do we have a much lower number of researchers, we send a far lower percent of our educated population to careers in research.

By these statistics, we see that India employs far fewer researchers than the United States and China. This is the generally cited reason for India's poor performance in research. It is also the opinion of Smriti Irani, former minister of Human Resource and Development, who once told the Parliament that India's poor research record is attributable to our small research cohort. But is this the whole story?

Productivity of Our Researchers

Our objective metric of the productivity of researchers is the number of citations that their publications garner. If a researcher's citation count is high, we may consider that his or her research is influential: meaning that the research has a high potential for disruption and for creating economic value and/or public benefit. Collectively, the citation counts for all researchers in a country comprise the total measurement of the country's research productivity.[5] How many of our researchers are highly productive?

Let us consider what proportion of our researchers are high producers, producing say more than 10 citations per year. We expand this notion to find what is called the "distribution" of research productivity. Rather than the single number 10, we identify a set of numbers and then calculate how many researchers achieve at or above each number: for example, more than 1, 2, 4, 8, 12, etc., citations per year. This is similar to counting how many cricketers have scored more than 10,000, 20,000 or 50,000 runs in one-day internationals.

We use Microsoft Academic Research data to do this analysis.[6] We base this exercise on the year 2013. We mapped 87,000 authors from India, 271,000 authors from China (3.12 times that of India), and 621,000 from the United States (7.15 times that of India).[7] One may note that the US-to-India ratio is similar to that reported above, but the China-to-India ratio is 3:1, compared to 6:1 reported in the previous section. This implies that the reported number of researchers in China is questionable. The number might be inflated or China might be counting individuals who perform research but do not produce papers. For our present analysis, we are concerned only with the distribution of authors across countries rather than with absolute numbers. Figure 4.1 shows the distribution by percent for the United States, China, the United Kingdom, and India.[8]

We observe that the United States and the United Kingdom have similar distributions. China has a weaker distribution but India scores the worst of the four. In India, the proportion of researchers doing high-quality research decays the fastest across the distribution. Approximately 55 percent of our researchers write papers that get no citations within two years of publication. In the United States and the United Kingdom, only 35–37 percent perform similarly, while the figure for China is 46 percent. Researchers who earn more than eight citations per year comprise 30 percent of the research community in the United States, 21 percent in China, but only

Figure 4.1: Percentage of Researchers Versus Number of Citations

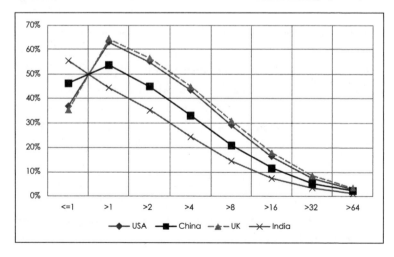

14 percent in India. Here, China performs 1.5 times better than India, while the United States performs twice as well. For truly disruptive research (>64 citations), China scores 1.8 times higher than India. The United States leads with 2.23 times our output.[9]

From these numbers, we can calculate the number of highly productive researchers. We multiply the total number of researchers in the country by the proportion that produces good research (>8 citations). By this calculation, the United States has 470,000 (1.4 million × 29.23 percent) high performing researchers, whereas India has only 27,550 (190,000 × 14.5 percent). The United States outperforms India by a factor of 17! China is somewhat more complicated due to the researcher count inflation. Their publishing researcher number must be somewhere between 3 and 6 of India. If we take 5 as a guess estimate, we can say that China has around 7.24 times the number of high performing researchers as India.

Simply put, the productivity of our researchers is much lower than it should be for a country with a fast growing economy, an educated population, and aspirations for global competitiveness. Not all researchers can be superstars for any country. However, a healthy ecosystem maintains a good ratio between total researchers and those that are high performing. Taking the United States and the United Kingdom as a standard, the proper ratio seems to be around 30 percent. India has less than half. We might as well admit that we effectively have half the number of researchers as we claim. Continuing in this manner, we will get one true researcher for every two that we add. This is neither economical nor rational. We need to concentrate our efforts and our investments in having researchers that can perform and produce according to world standards. Blindly and arbitrarily increasing the number of researchers is of little benefit.

Productivity is both a function of the ability of the researcher and of the environment in which he or she operates. Obviously, there is no absolute objective way to isolate the effect of these two. However, in this chapter, I will argue and bring evidence that we are not doing enough to attract the most advanced and proficient talent to research careers.

Now let us consider a particular slice of our researchers: the PhD students. Who are they, and why is it important to count them separately?

PhD: The Training Ground and the Workhorses

Piled Higher and Deeper.

—PhD as expanded by Jorge Cham,
the author of *PHD Comics*

The combination of faculty members and PhD students working together has proven to be a highly successful model of doing research. Wilhelm von Humboldt first proposed the idea of such research-oriented universities in the nineteenth century. He saw higher education as a holistic combination of research and academic study. He founded the University of Berlin in Germany as the first such institution based on his philosophy. Today, all top research universities run by this model.

A PhD student holds a master's or bachelor's degree and wishes to address an unsolved problem. He or she and a faculty member collaborate to solve this problem. The papers that they publish regarding their research endeavor measure the success of their effort. By way of this process, the student earns a degree higher than master's, called the Doctor of Philosophy, or PhD. A faculty committee bestows the diploma after deciding whether the research output is commensurate with the prestige of the degree. The requirements for the degree can take anywhere from four to seven years. After proving his or her ability through the academic and research process, the newly minted "doctor" should be capable of performing his or her own independent research work. Larry Page and Sergey Brin developed the technology behind Google as the result of their PhD research at Stanford, under their faculty adviser Terry Winograd.

The PhD students are the workhorses of research. In a model first pioneered in the United States, research faculty and PhD students work together as "equal partners in research" to bring out innovation. The PhD students join the faculty members as intellectual partners to conceive of research strategies, analyze results, and write papers. They also take responsibility for the "heavy lifting," that is, the actual operations and hard work required to solve the problem. They write the code, build the robots, work out the proofs, and conduct the laboratory experiments. The number of high-quality papers is strongly linked with number of high ability and hardworking students. Faculty guidance and student effort are equally important for research efficiency and success.

In choosing PhD students, Rosovsky writes, "We should also look for passion, preferably even an obsession with the proposed subject.... We should be looking for young men and women who find it difficult to distinguish between work and pleasure when it comes to academic tasks." I completely agree and search out such individuals for my own research team.

PhD students in top research schools lead demanding lives, like those of sportsmen. They work long hours, often in solitude, and take up multiple responsibilities; the challenges and demands that they face over the course of their research work takes an emotional toll. By virtue of their unique life, a cult has built up around them. They become like a fraternity, of which they are very proud. We can see examples of this association with the comic strip *PHD Comics* that satirize PhD life. Jorge Cham, a PhD student at Stanford University, first published *PHD Comics* in 1997, and it remains highly popular among graduate students around the world. I first read it during my days at MIT and could relate to a lot of its moments. People in my research group regularly circulate comics to our internal mailing list—many of them directed at me!

A second important aspect of the PhD pursuit is research training. Through his or her doctoral career, a student is taught how to follow the scientific process and produce something that is valid, defensible, and useful. The PhD process develops the research

Table 4.3: Total Number of PhDs That Graduate Each Year by Country (Data for Year 2013)

Discipline	India	Australia	China	US	India/China (%)	India/US (%)
Management	697	669	3,698	1,545	18.85	45.11
Medicine	1,472	1,219	8,228	10,976	17.89	13.41
Engineering	2,381	1,441	18,331	8,963	12.99	26.56
Agriculture	3,146	344	2,435	1,329	129.20	236.72
Others	3,215	581	3,809	5,945	84.41	54.08
Science	4,263	1,758	10,369	9,290	41.11	45.89
Social Science	7,893	2,084	6,269	14,712	125.91	53.65
Total	**23,067**	**8,096**	**53,139**	**52,760**	**43.41**	**43.72**

Source: India: Ministry of Human Resource and Development, USA: National Science Foundation, Australia: Department of Education and Training; and China: Statistical Yearbook.

manpower for a country. As the coursework of a Bachelor of Commerce degree trains a person to become an accountant, and an engineering program trains engineers, the PhD trains students to become researchers. The number of researchers for a country is a function of the number of PhD students that the country graduates each year.

Table 4.3 shows the total number of PhDs that graduate each year in India, China, the United States, and Australia.[10]

We find that India produces slightly less than half the number of PhDs that graduate from China and the United States. China and the US graduate almost equal numbers of PhDs. Among the various disciplines, India does extremely poor in medicine and engineering (graduating only 15–25 percent as many PhDs as the United States and China). As noted in the last chapter, both these fields are extremely important from the standpoint of economic returns and public good. Agriculture is the only field in which we graduate more PhDs than the United States and China. Still, India's rank in "Agriculture and Biology" is between 12 and 16. Even in this field, China outperforms India in research citations (12.3 percent and

2.7 percent of world citations, respectively). Although agriculture is important for a large agrarian economy such as India's, our share of agriculture PhDs is disproportionately high.

Not only do we have fewer researchers than the United States and China, we add fewer numbers than those countries each year. At the present rate, we cannot catch up. We do not have enough individuals undergoing research training. Given the importance of the faculty–student research model, the lack of workhorses, aka PhD students, impedes our productivity further. For a discussion on postdoctoral fellows, see "The Missing Postdoc."

The Missing Postdoc

Postdoctoral researchers (or postdocs) involve themselves in research after they complete their PhD degree. The postdoc position is mostly a transit stop between PhD study and faculty appointment. My lab in MIT had two postdocs—one had completed his PhD in our lab and was continuing as a postdoc for the year, while the other had done his PhD in Italy and was spending a year with us. I befriended one of them—he took me out for dinner as a welcome to the group, and this was followed by variety of intellectual discussions.

The postdoctoral position provides many opportunities in the research ecosystem. It helps a PhD student to build on top of his or her doctoral work by continuing with his or her professor. The professor continues to enjoy the benefit of a trained partner, without affecting the graduation time line of the student. The postdoc station also functions like a prolonged interview, so that professors and institutions can get to know a candidate before offering them a tenure-track position. The position also allows institutions to engage high performers while they pursue full-time opportunities elsewhere. Similarly, it affords the postdoc fellow the opportunity to interview the institution and richen his or her experience before joining the institution or moving on to other things. Some institutions even require postdoctoral work to qualify for faculty positions.

Unfortunately, the culture of postdocs did not develop well in India. We have very few postdoctoral researchers. Institutional positions for postdocs do exist, but they aren't widespread or very attractive. PhD graduates would rather move quickly to secure a faculty or industry position that offers better salary and status.

The missing postdoc presents a lost opportunity, for students, faculty, and institutions alike. Postdoctoral programs, if developed strategically, may offer many benefits. They can create an inflow of high achievers within our university, who are otherwise not seeking full-time careers with them. They can help improve the productivity of the research faculty. Some of them might even convert to faculty!

Having considered the quantity of our PhD students, now let us consider quality. Only 11 percent of PhD students in top Indian research institutions come from undergraduate programs at IITs and IISc, and 19 percent from top 50 colleges and IISERs. The remaining 70 percent come from rest of the colleges.

In contrast, of Indian PhD students in the United States, 56 percent graduated from IITs and IISc. Our higher research institutions do not attract our best and brightest. This is not to say that there is no talent other than undergraduates from IITs/IISc.[11] However, these top undergraduate programs are very effective at recognizing and recruiting our best high school students. The lion's share of talented undergraduates are in such schools. Our PhD programs cannot afford to miss on them.

Naturally, these quality issues are reflected in the productivity of our institutions. The proportion of researchers who do high-quality research in IISc is 14 percent according to Microsoft Academic Data (MAR data). The figures for Tsinghua University and MIT are 21 percent and 28 percent, respectively. China again outperforms India, this time by 1.5, while MIT matches twice our output. In the category of disruptive research, Tsinghua is almost twice better and MIT is three times our superior.

Also, we have fewer number of good research institutions. In the United States, there are many institutions of comparable fame to MIT, the perceived greatest: Stanford, UC Berkeley, and Caltech, for example. By contrast, India has few universities comparable to IISc. Thus, not only our top institutions attract fewer good undergraduates, they themselves are low in number, providing lesser avenues of PhD to our brightest.

India offers few opportunities for students who wish to pursue research training. India has few quality research institutions, and they attract few of our best undergraduates. As a result, their productivity is half that of graduates from the world's best universities.

Focus on *Quanlity*

India suffers from a low quantity of high-quality researchers. Or as one young Indian professor stated it to me, "Lack of critical mass of decent researchers in the field." We need our best to be in research. High-quality researchers produce research that has a greater impact on the field than the collective research of a large number of low-quality researchers—even if the collective citation count adds up to the same number. As we will see in the chapters ahead, high-quality researchers have a "network effect," making the productivity of all researchers higher. Think of this way— interaction between two high-quality researchers can help them both see their work in a new light, thereby producing better output than they would have produced on their own.

We require a shift in mind-set toward what I like to call *quanlity*, that is, the number of high productivity researchers. I calculate our present quanlity at 27,550: the number of researchers who produced papers with more than eight citations in 2003. This is poor. The United States has 17 times more highly productive researchers and China has eight times more. Similarly, our count and productivity of PhD students is much lower.

We need to determine what portion of our budget is improving our quanlity of researchers and what share is merely bloating our quantity of researchers. Presently we add around 25,000 PhD students each year. We need to ask how many of these will eventually supplement the 27,550 cohort. Judging by current statistics, probably few. To improve this, we need to get more talented individuals into research. That is the limiting factor in their performance. How do we do this?

In any nation, a set of complex cultural factors, institutional mechanisms, and incentive structures determines the careers that talented people choose. We can broadly define these factors as below:

1. Awareness, culture, and selling: As a nation, do we respect the profession of research? Are we aware of how scientific research contributes to human progress and our quality of life? Do our top college graduates understand the research career? Are they aware of what researchers do and of what Indian institutions can provide? What efforts have been put in place to spread awareness about research? Do our institutions respect prospective students and hard sell open positions in research programs? These are basic ingredients to attract individuals to research.

2. Personal and professional rewards: Do our undergraduate students recognize the rewards offered by a career in research? Personal rewards include salary and social status. Professional rewards are even more persuasive: highly intelligent people often seek opportunities for exercising their intellect, achieve excellence, and make impact. They look for work environments that will help them achieve their best. For instance, faculty members look for high-quality peers and students. These are much required for them to produce top quality research. Similarly, prospective PhD students care about the quality of the departmental faculty, one of whom will become their own faculty adviser. Can our institutions satisfy the desire for personal and professional rewards, thereby encouraging students to take up research careers?

3. Career path: Finally, people care about the potential for career growth and the range of options available. A student who is considering investing time and effort to obtain a doctoral degree would naturally like to know that he or she has rewarding choices for postgraduation. A faculty person would feel more energized if the academic world were diverse, with a variety of outlets for teaching, research leadership, planning, and administration. These rewards go beyond financial ones. For example, consider the number of academics who went to work for the Obama administration. Do our own research ecosystem and institutions offer an exciting and rich set of career paths?

To answer these questions, we take help of data from a survey of prospective PhD students and faculty members. We asked them what they think about research careers and the research ecosystem at home. We surveyed three sets of people: top engineering undergraduate students,[12] Indian PhD students in top institutions in the United States, and PhD students in top institutions in India. Our sample included 241 undergraduates, 86 PhD students in India, and 46 Indian PhD students in the United States. The Indian PhD sample came from institutions such as IISc, IISERs, TIFR, and the IITs. The US sample included PhD students from universities such as MIT, University of Michigan, Stanford, and Caltech.[13] We combined this data with supplemental public data to reach our conclusions. Let us start by understanding how our top students perceive a PhD degree from India.

Choosing to do PhD

Brain imbalance, not only brain drain.

I joined MIT's PhD program in Electrical Engineering and Computer Science in 2005. Since a young age, I have had a very curious mind, wanting an answer to every possible question. Many people shed their curiosity as they grow up, but my curiosity only deepened with age. It seemed only natural that I would enter a career in research. I applied to top four schools in the United States for my PhD and won admission from MIT only. I wondered if they had made a mistake—statistically plausible! Whatever the reason for my good fortune, I joined a cohort of some of the best students from around the world. My Indian colleagues came from the IITs, including one JEE (IIT entrance exam) topper and other department high achievers. Such environments can be intimidating. Why do high achieving kids put themselves in such stressful competition for elite PhD programs? Why do they not consider Indian institutions?

We asked final-year undergraduate students in engineering if they would like to pursue a PhD, and where. Their responses are tabulated in Table 4.4.

We found that less than 20 percent of our top undergraduates want to pursue a PhD in India. In the case of the top 50 engineering colleges, this number drops to 7 percent. About 46 percent of students want to pursue PhD abroad, while 35 percent wish to pursue a different career—a job or an entrepreneurship.[14] Please note how research loses students to industry and US migration in a 40:60 ratio. In our top educational institutions, five times more students opt for careers in industry than those who opt for research.

Many people cite "brain drain"—the outflow of our brightest minds—as the primary reason for our poor domestic talent pool. However, industry and entrepreneurship claim at least as much of

Table 4.4: Response Statistics—Would You Like to do a PhD?

	Total (%)	IITs/IISc (%)	Top 50 Eng. Colleges* (%)	Other Eng. Colleges (%)
Yes, from abroad	46.5	67.4	53.6	41.7
Yes, from India	18.3	4.7	7.2	25.7
No	35.3	27.9	39.2	32.6

Note: *These include IITs/IISc as well.

our best research potential (since 2015, entrepreneurship has been claiming an ever larger share of talent). A professor from IIT Delhi has told me that they get much better quality PhD applicants in a recession year when the job market is down.

This is not to say that industry and entrepreneurship do not create value for India. However, in the current scenario, there is a "brain imbalance." Without dedicated talent flowing into research to support industrial expansion and create opportunities for entrepreneurship, economic growth will ultimately stall. India must not only compete with research opportunities in the West but must also lure students away from alternative career paths in India. If we can design our PhD programs and recruitment in a way that makes apparent the real benefits of the research life, then we will make a good start.

In a follow-up, we asked the industry-inclined students for reasons about their lack of interest in research. Approximately 25 percent reported that they have no interest in research.[15] This set might not be completely lost; lack of interest is often a problem of culture, exposure, and awareness. Let us discuss the top three reasons that students give, beyond lack of interest. These reasons accounted for 50 percent of the student sample.

Students cite the long duration of PhD study and lack of good career paths postgraduation as the major deterrents for pursuing research careers. Students cited these reasons regardless of college rank. A PhD program can range from 4–6 years. Such commitment is not very attractive to students who dream of moving ahead fast in the contemporary accelerated corporate environment. In our increasingly fast moving world full of automation and convenience, youth tend to have shorter attention spans and want to do everything quickly. PhD programs have not adapted to these changes, while the industry has. Such commitment also deters people who need a supporting wage for their family. Students need to have a strong personal motivation for research in order to toil so long in the university.

An important reason cited by students outside the top colleges is lack of awareness. Approximately 25 percent of high achieving students enrolled in colleges other than the top ones said that they have never received guidance to steer them to research. By a similar percentage, students at top colleges cite the lower stipends available for research study. They would rather take advantage of the hot job market in IT and start-ups to earn good money while they can.[16]

Let us now look at why the West is more attractive. For those who are inclined to research, what would compel them to pursue their degrees at home rather than abroad?[17] This is another question we asked. Across all colleges, students cite the same four out of five factors: better faculty members, better infrastructure, a better research environment, and better career opportunities post-PhD study. It is worth noting that three of these fours are related to professional rewards during their study: enacting such changes would help them perform better professionally. We will examine each of these in detail in this book.

Finally, we examined the motivations of students who stated that they wished to pursue their PhD research at home.[18] These motivations can help us understand what is working for the Indian system currently and which efforts may be replicated and strengthened.[19]

The top two reasons for wanting to pursue a PhD in India are love of country and a desire to help address domestic issues. An overwhelming 63–77 percent of students cited these two reasons. These are also the reasons given to me by almost all young professors who return to India after their studies in the United States. Some of them reported that they could not clearly articulate their reason for returning: they just said that it seemed natural for them to live and work in India.

Love of country is the one key persuasion for high achieving people to stay in India and works toward national development. They are ready to forsake their personal and professional interests in the hope that they can beat the system to make a difference. This is a splendid silver lining to an otherwise dark cloud. However, this fervor alone will not change the research productivity in India nor induce sufficiently large numbers of high achievers to stay at home and work for change. Moral sensitivity can attract a few noble souls to India who are not motivated by material incentives or market

forces. But such people are few. Besides, a healthy research environment requires a diverse set of researchers with diverse motivations.
In fact, I would argue that we precisely need ambitious folks who
fiercely care for their career and high achievement. Such people
can upend the stagnant order and build a dynamic and competitive
environment that asserts itself in society, arguing forcefully for its
interests, and opening up new career paths and possibilities.

We find that India's present pace of economic development
and current public optimism also helps attract people. In the last
50 years, Indians have never been so hopeful. Good public policy
can help rub off some of this optimism to the research world.

We have very good insight into how students from various
colleges perceive research careers in India. We lose many of our best
students to research institutions in the United States and to the
Indian job market.

Let us further investigate the factors that influence peoples'
choices.

Awareness about Research and Researchers

Each year, Indian newspapers publish the names of the top scorers
on the CAT and UPSC Civil Service exam on their front page. The
CAT is the entrance exam for prestigious management programs like
the Indian Institutes of Management and others. The UPSC exams
are the gateway to the much-coveted Indian Administrative Services
and subsequent careers in government that bring considerable power,
respect, and responsibility. A massive multimillion dollar coaching
industry has grown up around these tests to prepare candidates for
exam day.

The CAT and UPSC need no introduction, but how many have
heard of the exam for entrance to prestigious PhD programs? Has
there been any mention in the newspapers or on TV? Does anyone
know the high scorers on the GATE or NET in recent years?[20]
Probably not.

Our top institutions recruit PhD students through a competitive exam called GATE and NET.[21] The lack of coverage around it is a testimony to how much Indian society values research careers relative to how they value careers in management, government, engineering, or medicine. Generally speaking, our society does not hold research careers in high regard. One might attribute this to a lack of awareness but we also must fault the poor example set by the low quality of research in India.

Students report a lack of awareness of and lack of guidance toward research careers in their top three reasons for not choosing to pursue such a career track. Even after students are informed about research careers, they might not choose to pursue it. However, our education system should provide students with thorough knowledge of all possible career paths for them to make an informed choice.

Our survey data supports the hypothesis of low research awareness. We asked students if they could name any three research journals or conferences. Approximately 70 percent of our top students could not name any. Even in the top institutions, more than 60 percent of students were unaware of research literature. This 70 percent is an opportunity lost. They have never had enough exposure to research to make an informed decision about whether or not to pursue it as a career.

Next, we asked students if they could remember reading a paper published in an international journal or conference. We also asked them, if possible, to give us the title of the paper and the journal/conference it was published in. The results are shown in Table 4.5.[22]

Table 4.5: Response Statistics—Have You Ever Read an International Journal or Conference Paper?

	Total (%)	Only IITs/ IISc (%)	Top 50 Eng. (%)	Rest Eng. (%)
Yes, one or more (Valid)	24.4	54.5	37.1	17.3
Did not read any	75.6	45.5	62.9	82.7

We found that only 37.1 percent of students from top institutions and 17.3 percent of students from other colleges had read a paper. On the other hand, more than 50 percent of students in the IITs/IISc have had some exposure to research.

Finally, we asked students who had indicated that they were interested in research if they could name the top three research institutions in India.

Most students are aware of the IITs. This is not surprising given that they are top ranked undergraduate universities with many academic disciplines, other than being good at research. However, only about 50 percent of students in lower ranked engineering colleges identified IISc as a top research institution in addition to IITs. This is despite media reports on IISc's high rank in research. This implies that 50 percent of students in non-elite colleges are unaware of IISc. The lower reporting of institutions such as IISER and TIFR is understandable given that their focus is more on science than technology, although TIFR has a good computer science and technology department as well.

Why is our students' awareness of research and research careers is so low? Let us reflect.

Sowing the Seeds of Research

In the United States, sometime in the late 1800s, the idea of the public intellectual took hold. Until then, common people looked to the clergy for guidance on various issues of public life. Gradually, society took a more rationalistic and scientific orientation. More often, authorities were learned "experts" who could base their opinions in empirical evidence and systematic research. Some contemporary examples of public intellectuals in the West are Noam Chomsky, Thomas Piketty, Richard Dawkins, and Stephen Hawking. Such people are consulted by public officials (the US government) to offer input on various topics of health, science, and economy, for example.

India is experiencing this change, albeit very slowly. Politicians, bureaucrats, journalists, and religious leaders all seem to have strong voice and even pronounce verdicts! When I think of public intellectuals in India, the best I can conjure are Abdul Kalam and Ramachandra Guha. We do not hear the voices of scientists and researchers in India. They do not play a significant role in our society. We do not know their names and we do not have role models. This lack of public recognition and public regard turns our students away from what could be rewarding careers of great benefit to themselves and to the nation.

When society fails, educational institutions and teachers must take responsibility. During my undergraduate career, Professor Raj Senani, the head of the Department of Electronics and Communication Engineering, often discussed research papers in our class. He was himself a practicing researcher; research papers became a natural part of our class discussions. With characteristic modesty, he once told us how a previous undergraduate like us proposed a better circuit than that which he had demonstrated in class. The student then developed this circuit and published a paper. Professor Senani was a great teacher, always prepared, organized, and patient. Not surprisingly, I published three papers as an undergraduate on analog electronics, a subject he had taught me. How many other students have the benefit of such teachers and exposure?

We surveyed the students on whether any of their professors had discussed research papers in their classes. We asked for the name of the faculty member and the topic of discussion. No such discussion took place for 75 percent of the students. In the IITs/IISc, only 50 percent of students reported that faculty members discussed research.[23] This stands in great contrast to institutions such as MIT or Stanford, where almost every faculty member naturally discusses the latest research in the field in undergraduate courses. Majority of faculty members in India are apparently unaware of research and cannot motivate students about research. Authors such as N. Jayaram have noted the diminishing quality of teachers in higher education. We have a related challenge: the absence of teaching assistants, or TAs. TAs are PhD students who help professors teach courses by running tutorials, helping students with homework assignments, grading exams, and fielding questions. With particularly large classes, several faculty instructors and TAs may together teach the course. In many universities, PhD students must assist in at least one course as part of their degree requirements. In the 300+ research universities in the United States, undergraduates gain exposure to the PhD career in the form of class TAs. They interact with them closely and regularly, see their daily routine, learn about their personal research projects, and thereby might aspire to become one. I am not sure how many Indian institutions offer this kind of direct exposure. Outside the IITs/IISc, the concept of TAs is pretty much unknown.

Institutions that run PhD programs have the greatest incentive to spread awareness of research and to sell the research life. They are the most invested stakeholders. They have incentive to "recruit" high-quality students to their programs and turn them into great successes. Today in society, there is a global war for talent among all top organizations: Google, Amazon, Baidu, etc., all are competing for the rare few. The same is true for research universities in the United States—they try to attract the best, from wherever they can find them.

In India, industry goes to great lengths to find and recruit high-quality talent. Company representatives visit several campuses and conduct placement drives to hire top students. They compete aggressively with each other to get the first shot at interviewing the best students. Their senior officials make presentations about the company to the students and advertise the personal and professional benefits they offer. Companies such as Infosys or Wipro visit over 200 campuses every year. Companies such as Google and Microsoft visit more than 30 top colleges. My colleagues and I also visited campuses including NSIT, IIT, DTU, and several NITs to pitch research engineer positions in our organization. Some companies sponsor competitions and other fun activities to better engage students, buy media space on websites such as naukri.com, and others. They put their best efforts into acquiring top talent.

How well do our top institutions convince students to dedicate themselves to the PhD research effort? Do university officials visit

Table 4.6: Response Statistics—During Placements at Your College, Did Any University/College Personnel Come to Pitch Their Master's or PhD Programs?

	Total (%)	Only IITs/ IISc (%)	Top 50 Eng. (%)	Rest Eng. (%)
Yes (Valid*)	5.5	20.9	8.2	3.8
No	94.5	79.1	91.8	96.2

Note: *We consider those responses valid which named a university.

colleges to pitch their MS/PhD programs? The answer is shown in Table 4.6.

While 11.4 percent students responded in the affirmative, only 5.5 percent students could name a university whose officials visited their campus to pitch a PhD program. Half of those who answered affirmatively indicated that the visiting institution was from abroad: Nanyang Technological University (NTU) and National University of Singapore (NUS; Singapore), University of Florida, Trinity University (Dublin), State University of New York, and even embassy representatives from France and Japan. NTU and NUS in particular are world-class research institutions. Visiting Indian institutions included the IISc, IIT (Guwahati), Lovely Professional University, Thapar University, and National Centre for Biological Sciences (NCBS). We also asked students if any external faculty came to give a talk about his or her research to their class. Only 8.3 percent of students reported that anyone had given a talk at their institution. For the top colleges, the number was slightly better, but only 14.4 percent. Furthermore, of the professors who visited, one-fourth were from foreign universities.

The outreach efforts of our research institutions are completely broken. They put no effort to attract top quality students to PhD programs. Most of our schools are passive and rely on GATE to supply them with PhD students. The newspapers run a small government announcement in their advertisements with no complimentary effort to encourage good students with high potential to take up a PhD career.

In contrast to our dismal efforts, the top five US research schools actively compete among themselves to find the best students from around the world. They organize open houses to sell their programs to prospective PhD students. They allocate alumni mentors and provide various fellowship programs. In fact, one does not even need a bachelor's degree to gain admittance to a PhD program at MIT. If a student can demonstrate that she is among the best in research, she could conceivably gain acceptance. These institutions disregard rules and formalities when searching for the best. They see their educational mission in a different way. Recently, all the top schools have begun offering some of their classes for free through online platforms such as Coursera and edX. These are attempts at democratizing education, but they are also active sourcing channels for these schools. The IITs also have a similar initiative called National Programme on Technology Enhanced Learning (NPTEL), but this hardly enjoys the popularity of Coursera (from Stanford) among our top students.

Professional Rewards

Of the top five changes that students would like to see in PhD research in India, three concern professional rewards. They include better faculty, a better research environment, and better infrastructure. How do our students perceive the professional rewards offered by our universities? Our universities may not offer the best, but do they even know what is out there on offer? For example, do they know what kind of research funding is available? Do they know about travel grants?

Students have a low opinion of the quality of faculty members and researchers in Indian institutions. Almost half of all students think that less than 20 percent of researchers in top institutions perform world-class research. More than 75 percent of students (80 percent in top institutions) think that less than 40 percent of researchers are exemplary. Only about 18 percent of students can name a professor who they think is world class (some mentioned

C. N. R. Rao and Manindra Agarwal, but others mentioned obscure names as well as non-Indian researchers.)

This shows that we lack role models whom students might like to follow and work with. Therefore, high achievers do not apply, and truly world-class faculty does not come because they think student interest is not there. It becomes a self-perpetuating cycle.

Travel funding is among the most important supports required for great research to happen. Although Asia sponsors some worthwhile events, most of the times top conferences take place in the West. Conferences are not merely forums for the presentation of papers. They give opportunities for scholars to meet and interact with other researchers, to share their knowledge, and learn from each other. However, foreign travel is expensive; researchers of any nation or institution typically require sponsorship to fund their attendance. The government has taken a number of steps to improve funds for research travel, but it still remains a challenge for many of our people. Travel funding is a great selling point for undergraduates as a benefit to pursuing research careers. Unfortunately, only 20 percent of students feel that there is reasonably sufficient funding available for it. A majority (57 percent) thinks that it is difficult to get. Moreover, a hefty 20 percent thinks that travel funding is not available at all.

There have been efforts to improve other professional rewards of PhD study. Recently, some industrial organizations and private individuals have donated money to universities to use as funds for research and student stipends and for the improvement of infrastructure. Infosys and the Tata Group have made several initiatives. Kris Gopalakrishnan and Narayana Murthy, the founders of Infosys, have also made large private donations to university research. Most recently, Gopalakrishnan pledged ₹250 crores to IISc for its Brain and Cognitive research center, and Infosys Foundation made a commitment of ₹24 crores to IIIT Delhi for an artificial intelligence center. Only a quarter of our students reported that they are aware of this sponsorship. Only about half of the students know that some companies sponsor PhD fellowships.

Our students perceive that our university research environment is poor. They feel that the faculty is subpar, funding is scarce, and that both government and industry largely ignore the field. Although they are not entirely wrong, it is unfortunate that they do not know the brighter side of the picture. Our institutions haven't put enough efforts to highlight role models, the improvement in various funding, and their research achievements. Students are unaware that amidst the gloom, many prominent public and private individuals and institutions have demonstrated some faith and optimism about the potential for quality research in India. This optimism has yet to trickle down to the students.

Personal Rewards and Career Paths

There is no right way to determine the correct salary for a job. By principles of business and economics, it is the minimum amount that will attract and retain the kind of talent that the job requires. Better talent demands higher compensation.

In 2015, the Indian government increased the PhD stipend to ₹25,000 (per month) for the holder of a bachelor's degree and ₹28,000 for someone with a master's degree. However, this fact has been poorly advertised. Almost half of all students surveyed (42.5 percent) think that the PhD stipend is less than ₹20,000. In US universities, the stipend ranges from $2,000 to $2,500 per month for science and technology programs. To be precise, MIT offers approximately $2,435 for a master's program and $2,664 post master's. From a purchasing power standpoint, this translates to ₹42,500 or 40 percent higher than the Indian stipend.

Table 4.7 shows how the PhD stipend compares with some industry salaries for entry-level personnel.[24]

The IT services industry recruits from among the top 15–20 percent of engineering students, whereas Microsoft/Google recruits from the top 3–5 percent of students. Ideally, our PhD programs should attract the type of candidate who goes to work for Google and Microsoft. However, our PhD stipend is around a quarter of what these jobs offer for entry-level positions.

Table 4.7: Comparison of PhD Stipend with Job Salaries

Profession/Program	Salary/Stipend
PhD	3 lac per annum
Software engineer–IT services companies (Infosys, TCS, Wipro, etc.)	3–3.5 lac per annum
Software engineer–IT product company, Microsoft	11.5 lac per annum
Software engineer–IT product company, Google	14 lac per annum

The PhD stipend cannot compete with industry salaries. Nowhere in the world does academia match industry in pay. MIT's stipend for a PhD student is probably half of what an average computer science graduate earns in the United States. However, the Indian scenario is different in a couple of ways. A PhD degree from a top institution in the United States is highly coveted. Students rarely leave a university such as MIT for a job, regardless of salary. The Indian PhD programs are nowhere near as prestigious. Therefore, they must compete actively with other opportunities in society. We must also consider that given the humble socioeconomic situation of many of India's students, salary is an important consideration. The recent hike in higher education fees adds to the financial burden of the brightest and most talented choose study over work. The need for a higher stipend was also highlighted in our survey results.

Personal rewards generally include other things such as social prestige and self-worth. However, Indian PhD programs fall short in these categories also. Admission to PhD programs is not highly competitive, and being a topper in the GATE or NET exam doesn't offer much gratification. In terms of monetary benefits, a PhD mostly earns much less after getting a job compared to an MBA or software engineer in a corporate setup. And since salary is frequently a mark of regard and respect, PhDs are in a sense at a disadvantage.

Some careers might have little total regard in society, but they might be endowed with a sense of internal self-worth. These might include music ensembles and artistic communities, particular religious groups, and communities that live according to strict political or social philosophies. Such groups imbue a sense of mission and act somewhat cult-like. In the United States, the PhD student community certainly does invoke a strong self-identity and corresponding self-worth. They even have a dedicated comic strip, comprehensible only to them. The Indian PhD community does not have such a tradition. In conversation, they betray a sense of themselves as second-class citizens in their institutions, subservient to their adviser and the institution. The institution "believes" that they have done a favor by inducting the student. By contrast, the US institutions actively compete to get the best PhD students. Faculty respects their students and consider them "equal partners in research."

Lastly, let us consider the career paths available to PhD graduates. Unfortunately, the seemingly enviable prefix of "Dr" does not translate to a plethora of career options. The most lucrative offers would be a private research lab such as IBM Research, Microsoft, or TCS Innovation. They offer upward of 14 lakhs (around $22,000) in salary and provide interesting research work to do. Such opportunities are few in number. The rest of the industry offers little. Indian companies are yet to climb the innovation ladder. Out of the 50 smartest companies listed by MIT Technology Review in 2016, none was from India. One hot area where the industry offers some positions is data science.

The other career option for PhDs is as faculty in institutes. We will find that the options in top institutions aren't very lucrative. On the other hand, the 20,000 other higher education institutions provide mostly teaching opportunities devoid of any research work. Given these options, it is not surprising that only a third of Indian PhD students want to stay in India to pursue a future career.

The career path dilemma is not unique to India. Doctoral students in the United States and China face challenges also. Opportunities change year to year as demand from industry and academia fluctuate. At present, however, demand from innovative technology-oriented companies in the United States for PhD graduates has never been higher. Google, Facebook, Tesla, and LinkedIn, to name a few, are fighting for them. Last year, Mark Zuckerberg himself appeared at a theoretical machine-learning conference (NIPS) to pitch his company to research faculty and PhD students. Naturally, some PhD areas are more in demand than others, but overall demand is high. Computer science in particular has seen a stunning rise.

Another difference between the US and Indian PhD experience is that US faculty feels responsible for placing their graduates in a good job. Professors introduce their students to their network of contacts, organize meetings, and write reference letters. Counseling the student in his or her career steps is a key element of the US graduate school system. The student's professors, personal adviser, department, and institution all make various efforts to alert the student to the various career options available. I recall seeing several announcements for career counseling sessions for senior PhD students at MIT.[25]

Now let us consider how desirable our PhD students perceive faculty careers in India. How much do our institutions care about recruiting the best faculty talent?

Research Faculty: What Motivates Them?

Soon we are dealing in specifics. I offer a high salary, inwardly cringing at equity considerations; add a most generous housing subsidy; throw in a small "slush fund"—sort of academic mad money; promise to help the spouse find a job and to get their one child into Cambridge's leading private school.

The lines quoted above are again from Dean Henry Rosovsky of Harvard University in convincing a young top-notch economist to join his department. Dean Rosovsky articulates the hiring policy for top research schools in the United States: ask who is the most qualified person in the world to fill the vacancy, and then try to convince that scholar to join. It seems so simple. However, the difficult part is coming to the realization that people are the essential ingredient of success. In research, they are the center of the innovation ecosystem. It is amazing what the dean is prepared to offer: a high salary, the promise of research money, and benefits for the family! This anecdote is from the 1990s. The war for talent has only intensified since then.

Do Indian universities follow the same philosophy and make the same effort? Do India's best PhD graduates aspire to become Indian faculty? Are our institutions attractive to international talent who wish to pursue teaching and research? What kind of benefits do our schools offer as bait?[26]

Motivation

Not many of my Indian friends from my PhD program at MIT have returned to India. A couple of them with PhDs in mechanical and civil engineering returned for family reasons. One female friend preferred life in India and so became a professor at IIT Mumbai. A fourth friend with a degree in operations research returned for three months but then went back to America. No others seriously considered India. Now they hold faculty positions or work in various research labs or companies in the United States. By contrast, many of my friends from MIT's business school have returned to try their luck in the booming Indian economy and start-up world.

According to our study, only 8.7 percent of Indian PhD students in the United States have immediate plans to return to India. Conversely, 43.5 percent wish to stay in America to pursue their careers. The remaining 47.8 percent suggest that they might return to India at a future time; however, as their lives and careers in America become settled, it is highly unlikely that they will ever come back. It is commonly known that a lot of Indians abroad have a genuine desire to return someday, but never make it. India could win them back sooner rather than later if the possibilities for research and career were more attractive.

For those who indicate any desire to return either now or later, their top two reasons are country and family/personal reasons. Nation and family can exert a strong pull, but they are not enough to convince a critical mass to come back. The next two reasons cited by approximately 30 percent of students is the ease of securing a faculty position and expectations for India's growing economy. The first of these warrants some discussion.

In some cases, people who are just not very competitive or desire a more laid-back lifestyle decide to come back. One professor who had done his PhD in the United States told me that if he had stayed in America, he would have been a small fish in big pond. By returning to India, he could be a big fish, given the few researchers who return, and the comparative prestige of a degree from a US institution. This mentality can be either good or bad. If the person perceives the Indian environment as a start-up environment, a launching pad from which to compete with the best in the world, then the idea is good. After all, this is what start-ups do: start small, innovate, and in time overtake the current best. On the other hand, if the person simply wishes to compete with Indian colleagues— sort of a one-eyed specimen among the blind—then the notion is bad. We do not need placeholders and seat warmers. If we want to inspire a new generation, then we need the best and most energetic to return and transform our research ecosystem into a world-class example.

Some people return to India because they are unable to get a faculty position in the United States. This may involve supply and demand reasons. For example, I am told that there are more open positions for theoretical computer science in good Indian institutions than in the United States. This lack of demand in the United States with little expectation of new funding and expansion convinced the person to return so that he or she could better pursue his or her interests. Machine learning is a fast growing field in the United States with a considerable number of new positions created annually. However, the supply of PhD graduates with this specialty outpaces the growth in faculty positions. Finding a faculty position in machine learning in an above average university in the United States is extremely difficult even for talented researchers. This presents an opportunity for India.

The top reason people cite for not coming back to India is the perception that the United States is the place to be for research. Students (around 50 percent) are deterred from returning to India by the condition of our national infrastructure and daily life. Unreliable electricity, air pollution, traffic congestion, public and private corruption, and sanitation put people off. Other reasons cited by students include the politics of Indian institutions and the bureaucratic culture. Inadequate salary is also a deterrent.

Despite the detractions of pursuing a career in India, students professed some willingness to return if a few things could be changed. The top three include: students would like to see a better peer environment and better students; enhanced research funding; and higher salaries. The first condition—better peers and students—was cited by 30 percent of students. This would be difficult to change. India cannot attract good researchers mainly because we do not have good researchers already. People wish to be among equals, and "like attracts like." Peers inspire each other and offer each other good advice and prosper. This is a catch-22 situation—we do not attract good researchers, because we do not have them. On the other hand, when peers begin to gather, an accumulation effect ensues. Then, good research institutions become better and better. My fondest memory of MIT is the intellectually stimulating discussions with great people wherever we found ourselves: in class, at lunch, or in a random campus lounge. Each interaction and chance meeting was a learning experience that generated new ideas. Researchers thrive in such communities and therefore seek them out.

Other students cited funding as an important prerequisite for coming back. One student wrote specifically: "Major government investment in my field (analogous to large research projects like Tsinghua and IOP in China)." One demanded an annual remuneration of 20 lac per annum to seriously consider India.

In all, less than 10 percent of Indian PhD students in the United States wish to come back to India. Those who wish to return are motivated by family considerations and a desire to serve India, not because they feel that India holds genuine substantive opportunities for them. Others wish to return because of the comparative ease of securing a faculty position. While some such individuals might be less than ideal, this motivation at least represents an opportunity for India to exploit the competitiveness in the US job market and attract highly motivated researchers back home.[27] If we can build a better peer community for researchers, this will have the knock-on effect of attracting better students, which potential returnees also wish to see. These things, coupled with higher funding for research and better salaries, will go far in creating a first class environment.

Awareness and Recruitment

Around 40 percent of Indian PhD students in the United States reported that their university or city received a visit from a top Indian institution advertising open faculty research positions. Some of the universities that sent representatives frequently: IIT Gandhinagar and IIT Hyderabad, and also one of the older IITs, IIT Bombay. IISc was scarcely mentioned. It is heartening to see that the new IITs are taking initiative to reach out to PhD students abroad. They have also been conducting interviews at various international locations.[28] However, our more established institutes are weak on outreach for recruitment.

Such low activity hardly represents an active interest in identifying and recruiting the "best talent" available. Probably a little better than the efforts to hire PhD students, we are mostly passive; our institutions prefer to wait for the right candidate to apply to them. Generally, we hear the recruitment process is improving, particularly in terms of decision time. However, complaints of politics, bureaucracy, and lack of respect continue to surface. We can observe an absurd phenomenon of insecurity in many of our best schools. The current faculty feels threatened of new high achieving faculty rather than welcoming them. An IIT professor told one applicant from a top US school that his work was of bachelor's standard and not the degree standard as defined by the IITs.[29] The IIT offered him the position and he accepted, but he thought for a very long time. Faculty members know that they do not compare on a global standard, but living in their "well" they want to believe and maintain the illusion that they are first rate (see "Vivekananda's Frog in the Well Story"). This perpetuates mediocrity and the complete opposite of the accumulation effect that make the US universities tick. We need a peer environment that is dedicated to quality and results, not territorialism and pettiness.

Vivekananda's Frog in the Well Story

This is an excerpt from Vivekananda's second lecture "Why We Disagree" at the World Parliament of Religions, 1893.

A frog lived in a well. It had lived there for a long time. It was born there and brought up there, and yet was a little, small frog. Of course, the evolutionists were not there then to tell us whether the frog lost its eyes or not, but, for our story's sake, we must take it for granted that it had its eyes, and that it every day cleansed the water of all the worms and bacilli that lived in it with an energy that would do credit to our modern bacteriologists. In this way, it went on and became a little sleek and fat. Well, one day another frog that lived in the sea came and fell into the well.

"Where are you from?"

"I am from the sea."

"The sea! How big is that? Is it as big as my well?" and he took a leap from one side of the well to the other.

"My friend," said the frog of the sea, "how do you compare the sea with your little well?"

Then the frog took another leap and asked, "Is your sea so big?"

"What nonsense you speak, to compare the sea with your well!"

"Well, then," said the frog of the well, "nothing can be bigger than my well; there can be nothing bigger than this; this fellow is a liar, so turn him out."

Author: Indian science and science institutions need to get out of their well.

Personal Rewards

Let us start with salary. Low salary is a key factor deterring quality talent from faculty careers. Table 4.8 compares the entry-level salary for research faculty at top Indian institutions to entry-level salaries in industry.

Table 4.8: Comparison of Faculty Salaries with Job Salaries, by Country

Designation	Annual Salary			Ratio of Salary of Assistant Professor with Others		
	India (INR)	China (USD)	USA (USD)	India	China	US
Assistant Professor	1,050,000	3,780,000	144,000	1	1	1
Microsoft Research Scientist (For PhDs)	1,400,000	3,800,000	144,000	0.75	0.99	1
IBM Research Scientist (For PhDs)	1,800,000	–	140,000	0.58	–	1.03
Google Fresher (For Bachelor Degree holders)	1,400,000	–	125,000	0.75	–	1.15
Microsoft Fresher (For Bachelor Degree holders)	1,150,000	2,020,000	104,000	0.91	1.87	1.38

Source: Compiled from various online sources including payscale.com, glassdoor.com, IIT website, faculty-salaries.startclass.com (accessed in December 2016).

Note: The government of India prescribes a salary band for assistant professors. Based on this band, the top institutions pay a gross annual salary ranging from 9 lacs to 11.5 lacs. We have taken a middle figure of 10.5 lacs for our comparison. Whereas there are stipulations for three or more years of experience for such salary bands, they are often waived. Blank cells represent the unavailability of relevant data.

Salaries for research faculty are less than what Google pays to engineers with undergraduate degrees. Microsoft pays a slightly higher salary to its new hires. In the United States, the reverse is true: salaries for assistant professors are 15–38 percent higher than these same companies. We should not be surprised that a great engineering student in India would choose an industry job rather than go for an academic career. After earning a low stipend and exerting great effort for 4–5 years in a PhD program, a great student will graduate to a faculty job earning less than what she could have made immediately after her undergrad studies.

Indian university salaries fall into the range of 60–75 percent of what industry offers for research. By contrast, US universities approximate entry-level compensation for industrial research (PhD-level) positions at companies such as IBM Research and Microsoft Research. Although US faculty positions are highly coveted for their prestige, they still compete well with industry on salary. On the other hand, university positions in India are neither prestigious nor lucrative.

The government has been taking notice of compensation issues and has introduced some programs such as the Ramanujan Fellowship and the Ramalingaswami Re-Entry Scheme. Furthermore, the government, PSUs, and industry have begun sponsoring new faculty chair positions. Some of these efforts are too little too late and others have become obsolete. Besides, awareness of these programs is fairly low among Indian PhD students abroad. Only 17 percent indicated that they were aware of the re-entry schemes, and only a quarter knew about open faculty chair positions.

A final word on salary. Money is a bad word in Indian institutions, unfortunately. Some philanthropists declare that they do not want their money to go to research salaries. Older faculty members and university administrators look down on measures that prioritize salary and other benefits. They say that they themselves came back for love of country and endured all hardships. They expect the new personnel to follow the same path and resist any urge for change. Such a mind-set does not help. Economic incentives are necessary, and ability must be compensated adequately.

There are personal considerations other than salary. For example, there is the "double body" problem: the spouse of the faculty recruit also needs a job. This situation is acknowledged in the United States, and universities work actively to help researchers find jobs for their spouses. The wife of one of my colleagues, a very good robotics researcher, needed a job. A top US university offered her a position and then worked with my colleague to find him a job in the university as well—on merit, of course. The poor guy was often at the center of our jokes for how he secured his big position by way of his wife's small job!

Furthermore, faculty require help with their kids. If both are busy researchers, then they need adequate care facilities. Many universities have day care centers on campus. Each day, I encountered the pleasant sight of little kids queuing up around 5 pm in Stata Center, the new building for the Computer Science and AI Lab at MIT, waiting for their scholar parents. This was a great selling point for CSAIL!

Indian institutions are yet to consider these finer issues. Institutions in regions such as Kanpur, Kharagpur, and Powai do not provide the best career options for spouses or the best schools for children. Many candidates prefer Delhi and Bangalore, where it is easier to address these problems. One big advantage our institutions provide is faculty housing on campus. Mostly, these are nice cozy homes amid attractive green spaces. However, the system is plagued by shortages and difficult bureaucracies control allocation. Faculty often must stay long term in temporary accommodations. Recall that 50 percent of Indian PhD students cite poor infrastructure and day-to-day problems as a major impediment. The universities need to create functional but a humble and sensitive world inside their premises.

No university can completely solve all these issues of personal rewards. However, they can demonstrate sensitivity toward them and make institutional efforts to address them. Such efforts will not only provide better personal incentives to researchers but also make them feel desired, worthy, and respected. If we need the best in the world, we need to make them feel welcome.

Professional Rewards Career Path

Our institutions do not offer great professional reward. However, there have been some recent attempts at improvement. One new measure is a start-up grant of 25 lacs provided to new faculty members. This allowance helps supplement research funding for new faculty and also provides start-up money to equip new laboratories. Additionally, chair positions and re-entry schemes provide new research funds. Unfortunately, few people are aware of these measures. The start-up grants are known to about 41 percent of students, whereas only 17–23 percent are aware of new chair positions and re-entry schemes.

Researchers desire an illustrious career by making a broad impact inside their own institution, among their peers, and also in the world at large. They do so by working on impactful projects that seek to solve big problems. My adviser at MIT was fond of saying that MIT is one of the few places in the world where it is more difficult to become famous inside compared to outside. Indian institutions work in the opposite way. They have a few super-motivated researchers who carve a name for themselves inside the institution, but are vaguely known to the world outside.

To enable our researchers to make greater societal impact, we must exhibit a culture of respect for science and meritocracy. People need to be trusted with large projects and money. Governments should involve top scientists in their work. As noted earlier, US officials regularly engage experts in government projects and sponsor public–private partnerships for scientific research. President Obama maintained the Global Development Council that included several university faculty such as Esther Duflo, MIT Professor of Economics. Other prominent faculty are involved in various governmental energy initiatives. Anant Agarwal, another MIT professor, runs edX, a $60 million public–private initiative that makes top university education open to all. In fact, the United States involves many top academic researchers in well-funded government projects. This also occurs in India, but usually in the form of government education and research-planning initiatives. For example, one of our celebrated scientists, K. VijayRaghavan, is the secretary of the Department of Biotechnology: a position reserved for bureaucrats.

I would like to end where I started. India has great number of people with an aptitude for research. They have made great contributions in the past and continue to achieve big in other parts of the world. If we put our act together, we will be surely be able to attract a critical mass of them to our research institutions. The time is now.

5

Resources for Research: Speed, Accessibility, and Merit

Without hands and feet He moves and grasps; without eyes He sees; without ears He hears. He knows whatever is to be known and of Him there is no knower.
—*Shvetashvatara Upanishad* on consciousness

Imagine that you are a professor of brain and cognitive sciences, and your research agenda is to develop advanced prosthetic systems for the seriously disabled. You experiment with ways to help people who have lost the use of their limbs and lost their eyesight—to help them see, touch, and manipulate the world. We know that the brain is the seat of experience. The brain talks to the conscious being—still an unknown phenomena. It relays experiences from the sensory organs to the conscious being and conveys instructions back to our muscles, joints, and sensory organs. We can construct sensors and robotic arms and fingers to receive messages. But more importantly we need to learn how to "talk" to the brain so that it listens to the messages we send, and we need to learn how to "listen" to the brain so that we understand what it is telling us to do. If you are unusually successful in your research, you might even be able to create the human brain (including the conscious being) alone in a test tube, capable of experiencing and manipulating the world.

First things first. Before you begin, you will need to set up your lab. Perhaps you want to begin with prosthetic arms. What resources will you need to build robotic arms that can sense, touch, and connect to human nerves? You can get a robotic arm "off the shelf," or you can design a new one from scratch. You can build the different parts with materials such as silicone, a 3-D printer, and a 3-D design software. In addition, you will need touch sensors, motors, actuators, control system circuitry, and so on for the electrical and mechanical control of the arm. You will need administrative staff to locate the items and handle the ordering, payment, and delivery processes. To assist you with the technicalities of mechanical and electrical engineering, you may need to bring in other faculty or PhD students with experience in the field.

Next you will require instrumentation that can identify brain activity, such as MRI, EEG, or MEG machines. Such machines are obviously quite expensive but they are indispensable if you wish to learn to talk to the brain. You are going to be asking human subjects to imagine touch sensations to different parts of their hands to see which part of the brain becomes active. You cannot do this without brain-imaging equipment. Then you will need technical staff to maintain these instruments properly, keep them well calibrated, and help to operate them. Once you obtain the images from these devices, the images must be interpreted by advanced computer algorithms. Therefore, you will need computers—probably a cluster of them—to perform large computations. You may even need the assistance of a computer scientist.

Once the brain-imaging and computer equipment tell you which parts of the brain light up when the patient imagines or perceives touch, you will need to get to work on trying to recreate this process, by artificially stimulating those parts of the brain. For this, you will need invasive electrodes that go inside the brain (!) or some types of noninvasive patches. These things will connect to the touch sensors in your artificial arm, so that when the sensors are touched, they will activate (hopefully!) the same areas of the brain as a real touch. Sounds simple? It's not. You will require substantial engineering, calibration, and iterations (trial and error with various parts) for all these parts to work together, while also taking care that the process does not harm the patient in any way. For the latter, you will require the participation of medical doctors for their advice and oversight.

When all the equipment and personnel are in place, how do you actually go about doing what you wish to do? You gain insight from previous work in the field published in journals and conference articles. However, these articles do not contain substantial engineering details, descriptions of challenges, nuances, and fixes. For the most part, they simply concentrate on telling a linear story of success. So in order to expedite your experiment, you may wish to invite a person who recently worked in this area to give a talk. For example, Robert Gaunt from the University of Pittsburgh recently published an article about detecting touch. He would make a great speaker and collaborator. What can you offer him in return? You will need to produce some incentive that helps him progress his research in return. And you will need someone to coordinate his visit, help with accommodation, and travel and talk logistics. So now you might need a larger administrative staff to manage all of these arrangements.

The next piece is to listen to the brain—how does it instruct the arm to make a movement? If the nerves of the person in the arm aren't damaged, you may simply read the signal there. If they are, you will need to learn to listen directly from the brain. This is similar to what we already discussed. You will ask the human subject "to intend" to move his or her hand and see which part of the brain is giving this instruction. You will then directly read this signal through say electrodes and send a message to the motors and levers in the artificial arm.

As your research delivers new, noteworthy, or even groundbreaking results, you will want to present those results at conferences and journals in order to enlighten other researchers. A multidisciplinary project such as this—involving 3–5 PhD students and 2–4 faculty members—will likely produce several papers over multiple years. Everyone involved will want to attend conferences and present some aspect of their work. They will also like to learn from other researchers about what is new in the field, use it to further their research, collaborate, and innovate further. You are going to need money for travel, lodging, and conference registration expenses.

And after your project meets with great success, you will embark on your next project. Having helped people with missing or paralyzed limbs, now you might like to help the blind to see. What will be next? Sky is the limit. You will pursue the fabrication of every sense and action, short of the brain itself with its sentient presence—until ultimately you decide to tackle that as well.

This is a typical example of a challenging modern multidisciplinary research project. It highlights all the resources a passionate researcher and his or her students require to achieve great research results. Let us list them:

1. Equipment: This includes large investments such as fMRI and MEG machines, as well as others such as 3-D printers, motors, actuators, electronic components, and computers.
2. Technical staff: Trained technicians who can operate and maintain the equipment and components.
3. Administrative staff: Staff to help with procurement of resources, financial management, facilities management, travel planning, documentation, and other various administrative processes.
4. Funds: For sponsoring research students, faculty collaborators, and all the above.
5. Travel resources: For visiting conferences, universities, and for bringing experts to your lab.
6. Journals/conference subscriptions: To gather information relevant to your work.

Do our researchers have these things? How well do our institutions support our researchers?

Equipment and Skill
Building Blocks

Researchers, particularly experimentalists, require a variety of hardware for their research. They also require the "skill" for assembling the components, operating the equipment correctly, and also for maintaining it. The above example requires the use of fMRI, EEG, and MEG machines in order to read what the brain is doing. You may know that these are the same machines that are used to diagnose brain ailments such as tumors or clots. A single machine like this could cost anywhere between ₹1 crore and ₹4 crore ($150,000–$600,000). For use in research, one might require an especially sophisticated and expensive machine that is capable of higher resolution. The machine might also require further calibration or modification to make it compatible with the needs of the experiment. And the machine requires trained staff to operate it. These are examples of equipment that a researcher could buy off the shelf or customize for use.

But equipment such as prosthetic arms are different. For this type of equipment, a researcher would typically be more closely involved in the development. The researcher might begin with an off-the-shelf prosthetic and then modify it. Or she might build it himself or herself completely from scratch. In some cases, the design and construction of the artificial limb might in itself constitute new research. He or she might need to design new touch sensors or a new motor or chip for controlling the piece. He or she might need to modify the mechanical or electrical system for the success of his or her particular application. Or he or she might need to replicate an artificial limb that has already been invented but not available off the shelf.

This requires two things: the ready availability of equipment and components at a reasonable price, and the expertise for building the hardware. The latter is not trivial. The replication of an already-known process can take months or years, and there is no guarantee that the replication will succeed. Replication requires a knack for mechanical design, experience in performing each task, plus the ability to troubleshoot problems and iterate. During my stint at MIT,

I was situated in a humanoid robotics lab. There, researchers took several years to build their robots—some new and some replicas. This was the case even though the design of the robot itself would only constitute a small part of their research. The algorithms they ran on were important. One famous alumna of the group, Cynthia Breazeal, now a faculty at MIT Media Labs, took nine years to complete her PhD: much of her time was consumed by building robots.

Consider another example. In microfluidics, designing hardware is an intrinsic part of the research. Consider the glucose meter, a machine used by diabetics to monitor their blood sugar. The blood flows through micro channels in a nail-sized chip, undergoing chemical reactions in order to diagnose the level of sugar in it. Ten years ago, these blood tests were done on large non-portable equipment. Now they can be done on a tiny device by the user himself or herself, anywhere, in fraction of the time, and with a smaller blood sample. Microfluidic research is about building new chip designs that can perform new analyses. Although the fabrication of microfluidics chips are already known, modifying them to specific purposes and fabricating them requires considerable skill.

In some fields, hardware is not so essential. A mathematician requires little more than a room and peace. Some math researchers find computers to be helpful, but in my experience, most despise them. Computer scientists mostly need computers, albeit fast and powerful ones in order to run complex algorithms on large quantities of data.

Social scientists traditionally did not need equipment for their work. Historians and anthropologists required support to visit libraries, archives, and repositories to view historical documents and artifacts, and also to meet and interview people. Economists spent their time working out equations, building theoretical models, and analyzing publicly available data sets. This has changed across many fields. Now there is a larger emphasis on empirical studies involving controlled trials. This often requires large-scale data collection, which in turn often requires collaboration with other diverse organizations, funds, help of surveying agencies, and sometimes equipment too.

For instance, consider a study on the impact of white roof paint on house temperature, people health, and well-being in slums. This requires cooperation with an NGO, which works with slum people, surveying agencies and sophisticated temperature-measuring devices.

Equipment Without Staff

Water, water, everywhere, nor any drop to drink.

—Samuel Taylor Coleridge,
"The Rime of the Ancient Mariner"

How well does the Indian research ecosystem provide hardware availability and skill?

Do Indian institutions typically feature off-the-shelf equipment such as fMRI or MEG machines? Quite a few medical facilities have an fMRI machine. However, only about 5–15 research institutions have them. In the entire country, there are only a couple MEG machines. These are relatively new. But in the United States, all brain research centers have them. The number is at least 100. Mashelkar often cites the lack of facilities, such as, those for superconductivity experiments at extreme pressure and state-of-art synchrotron facility (x-ray research).[1] Since the equipment is not available in India, we are unable to lead research in these areas.

The current availability of such sophisticated equipment in Indian research institutions is at a moderate level, neither nonexistent nor in abundance. One may note that the number of potential users of this equipment—high-quality researchers—is also small. There is no point in having such machines if there are few smart people who can put them to use. In this sense, the low presence (as opposed to the availability) of machines doesn't create a bottleneck today. However, it will become a bottleneck as we add more high-quality researchers.

These expensive machines are usually procured at an institutional or departmental level and shared across faculty. Procuring a new machine could involve a battle lasting several years, which we will discuss in the next section. Even with the machines available, teachers' access to the machines and the staff to operate them present real issues.

Faculty rates availability of sufficient and capable staff as the second lowest among 11 parameters of quality (50 percent faculty rate it below average). Technical staff is insufficient and doesn't have proper training to operate the equipment. They lack exposure to modern technology and equipment. This is both due to low salary of staff and an aging population of staff at institutions. Due to insufficient training of staff, the faculty needs to handhold them through the process. This creates an additional burden on the faculty. It also delays the project. One faculty member noted in our survey:

> We have really hard time getting skilled and trained staff. Most of them have degree, but no proper exposure to any kind of instrument. Even if they know to use it, fundamental understanding of the working of the equipment is missing. There is a need for lot of follow-up after giving the job instruction. Things don't get done as per the work flow and needs a considerable amount of personal interaction to get the things done on time.

In absence of good staff, the faculty members often rely on the PhD students to run the machines. In some institutions, many staff positions are part-time. Such staff work intermittently and in irregular schedules. There is little incentive for faculty to train them, or for them to learn new things. Other times, the staff is simply not willing to cooperate. A faculty at one of the IITs told me that they had to walk across town to use the institutional facilities because their own departmental staff simply would not cooperate.

Institutional politics and bureaucracy also interfere with the availability of equipment. Lack of staff often becomes an excuse to deny access to the machines. They refuse permission to faculty and students under the pretense that they do not want the machines to get damaged by use from amateurs. A professor of neuroscience at an Indian institute told me that he was unable to use a machine in a partner institute, because they did not think he was trained to use it. So the machine just sat idle!

By contrast, MIT has a formal process for gaining access to equipment at departmental and institutional facilities. The student writes a one-page proposal on why he or she wants to use the machine. These proposals are judged on merit and usually approved swiftly. A full-time staff member trains the student and he or she gains card-based access to the lab, allowing him or her to come and go as he or she pleases.

So while we do enjoy some presence of expensive ready-made machinery, the larger problem involves access. Lack of trained staff and lack of cooperation by university officials limit the usefulness of this equipment. Also disruptive is the slow procedure for procuring new equipment. Let us understand this now.

Blocked Blocks

While large machines may be present, specific projects demand procurement of varied and smaller components. Remember our example of assembling a prosthetic limb. Our researchers rate time taken/ bureaucracy in getting new hardware or funds, as the lowest among 11 parameters of quality (61 percent faculty rate it below average).[2]

The hardware necessary for building such a prosthesis is generally not available locally. Such components have been mostly engineered in the West and are still manufactured there. India does not have enough researchers who are buying such things; therefore, there is no vendor ecosystem to provide them. Thanks to the Internet, most things can be ordered online. But consider the shipping and customs delays. A researcher in the United States can receive a component from a supplier in as little as 24 hours. The same component might take a month or more to reach the researcher's desk in India.[3]

The shipping delay isn't the real problem. The procurement process for components and equipment is as bureaucratic and difficult. Each institution sets a price barrier for components, ranging from $1,000 to $5,000. Anything above this limit needs to go through a tender process and gain approval from the institution's procurement department. This is despite the fact that you may already have an approved grant in which you have budgeted for the hardware.

Take, for instance, the case of one of the IITs. Only purchases below ₹50,000 can be made directly by the faculty. Anything above this limit requires approval from a committee consisting of six faculty members. These faculty members put together a tender process and decide on the specifications. Then they must post the advertisement in a national newspaper or send a letter of inquiry to 8–10 firms. The requirement also must be displayed on the institute's website and notice boards. They must collect at least three quotes. After a choice has been made, an internal audit committee will preaudit the tender. Then, after the vendor signs the necessary documents of governmental conditions, the researcher can place the order. Each of these steps takes 15–20 days. The entire process takes about three months and could extend to six months.

In addition to the laborious paperwork and the constant running around between people and departments, you often must contend with hostility from faculty members and administrative staff with some casting aspersions on the need for such expensive equipment.[4]

This makes it infeasible to build challenging new stuff that requires trial and errors. You iterate in such a process. The components you use first may not provide the required functionality and you switch to trying other alternatives. In such cases, the long purchasing cycles can turn into a nightmare. Dealing with iterations can render a project infeasible. For this reason, many researchers in India purposefully devise a research agenda that does not require such a process. Science loses in the process!

In contrast, a student in any top US research institution can order a $2,000 instrument by a click of a button after he or she receives an e-mail approval from his or her faculty advisor. From the time that he or she decides, he or she needs it to the time that he or she has it in his or her hands: about one week. A public institution in the United States might have slightly more cumbersome rules. However, the administrative staff is cooperative and helps you through the system. In most cases, you receive your hardware within a month.

Access to equipment and materials is not the only thing amiss. The skill to use the components to build desirable products is lacking as well. For example, fabricating a microfluidics chip requires precision molding. Vendors in India have the equipment to do this. However, they have neither formal training nor experience in precision molding of chips. They use coarse molding techniques for other more simple purposes. In case of microfluidic chips, they do not get it right the first time and need several iterations. Much of the reason is that the lack of critical mass and demand of chips has led to an immature vendor network. In contrast, vendors in China do "get it right the first time." The industry has invested in the vendors. They have experience in doing it on a large scale.

Even though the process of fabricating microfluidic chips is standard, Indian PhD students struggle with it because of the lack of vendor skill. As we saw earlier, institutions do not generally invest in trained staff to compensate for poor vendor performance. Therefore, PhD students end up wasting a considerable part of their PhD in learning these processes ad hoc. Chinese and American PhD students are free of these obstacles and are able to spend their time on the critical parts of new designs, rather than in making up for manufacturing flaws.

Finally, we must acknowledge that building stuff is not always considered research in India. Although this question is debated throughout the research world, India in particular looks down upon parts construction. Building stuff might not be research in many cases, but it remains an integral part of the research process. The stuff that gets put together often forms the backbone of the whole research program. We will discuss this in more detail in Chapter 8.

In summary, Indian researchers encounter many obstacles in building hardware for their research. Local unavailability is the first barrier, followed by the overly bureaucratic procurement process. The skill for manufacturing and assembling stuff is immature due to an overall lack of demand for these skills, lack of institutional mechanisms to develop such skills, and the lack of respect for the building process itself.

Travel Resources: Why So Hard?

In 2003, I wrote a paper on designing a genetic algorithm for automatically developing new oscillator circuits. The paper got accepted at the NASA/DoD Conference on Evolvable Hardware, which was to take place in Chicago in the United States. Since the genetic algorithm was designed to solve a problem we had discussed in our circuit class, it was original research and not a derivative of an already-known work.

If I wanted to attend, I had to finance the trip myself: airfare, accommodation, local travel, the visa, and also the conference registration fee. My father immediately offered to give me the money, but I was not keen to accept. I was also bothered by an ethical question: how others with a more humble background would finance such a trip toward a scientific endeavor.

I learned that there was no institutional or government program for funding travel for undergraduate research. Faculty and PhD students get it, but not undergrads. My department head was unflinching in his criticism, lamenting "Is it a mistake on the part of an undergraduate student to do good research?" I wrote an application to the director and dean of the institute. After multiple follow-ups, I was told that my application will be decided by the Board of Directors at NSIT. I wrote another application to the national DST including a strong recommendation letter from my department head. This was turned down. I researched other possible funding organizations, but nothing worked.

When nothing worked, I reluctantly accepted my father's offer. My moral qualms remained, but as for the money, my father convinced me to view it as a loan for "investment" in a good thing. The conference folks were kind enough to find me a shared room with another student who had a paper. The conference hosts had fewer procedures than my own institute and country!

Faculty rates availability of travel money as the fourth lowest among 11 parameters of quality. More than half of the faculty members I talk to rate travel as their number one challenge. Around 57 percent of PhD students in top Indian institutions rate travel funding as poor or average (on a 5-point scale from poor to excellent). Among 11 parameters of PhD quality that we rated, travel funding is rated among the bottom three by PhD students. As an undergrad, I was told that there were institutional structures to support faculty and PhD students for conference travel. Why is it viewed as a major challenge for faculty members and PhD students?

To understand this further, let us look at a typical travel policy dictated by top Indian institutions.

Faculty is allocated a fixed amount (e.g., ₹3 lacs or $40,000) for three years for international travel. Additionally, they can apply for international travel funding from DST, but only once every three years. There may or may not be budget restrictions for individual trips. PhD students can get funding from their institutions to attend one international conference during their PhD career, up to a certain amount, usually around ₹1 lac. Faculty cannot use grant money from government institutions such as DST for travel. There is no institutional mechanism to get funding to invite international speakers or to visit foreign universities.

Notes: 1. These numbers may vary across institutions. Some provide 6 lac for three years and may consider a second trip for a PhD student. Also, some institutions may impose additional budgetary restrictions for each trip.
2. US conference trip cost: ₹1.5–2 lacs; Europe conference trip cost: ₹1–1.5 lacs.

This level of funding for PhD students falls short. A PhD student in the computer science department of IISc wrote the following in our survey:

Better funding for travel is required. After every publication, there is a big headache to arrange money for travelling. Institute funding is limited to ₹1 Lakh for every PhD students. While, requirement to finish PhD is 3–4 International papers! I think there is a huge divide in the expectation and facilitation of this effort. International conference typically happen in USA or EUROPE which leads to heavy expenditure (sometimes from own pocket).

Any competent PhD student writes 2–3 international conference papers during his or her PhD study. In my short two-year stint at MIT, I visited two conferences and one workshop in 2006 and 2007 each, for a total of six trips. I traveled to Istanbul, Seattle, Michigan, London, Nice (France), and Berkeley. I presented a paper during each of these trips (except for Berkeley).

Consider the travel difficulties of a high performing faculty member—our brain and cognitive science professor, for example—with five PhD students. Assume that each student writes two international conference papers in their three years, for a total of 10 across the PhD cohort. Faculty generally makes two trips a year: one for a conference they regularly visit and another to present an accepted paper, probably overlapping with a student. Thus, the professor and students' total count for 3 years approximates to 16 trips. The funding described previously would cover only half!

There are other issues with funding, other than the amount. One is the bureaucracy, as mentioned by our student commentator. Even when funds are available for a second trip for a PhD student— perhaps through alumni sponsorship—the student must seek approval from the director of the institute. An American academic would find it crazy that the director of an organization would busy himself or herself with travel requests from students. Anyway, being unfamiliar with either the student or the student's research, the director cannot be considered the best judge the worthiness of the request. Such a system as this induces bureaucracies to add layer upon layer of requirements, forcing you to run around for stamps and signatures when you should be preparing for your event.

We also tend to be "socialist"—theoretically the institution wishes to spread the money to provide more students and professors with opportunities, not just the high performing ones. While laudable in its ideal, it is difficult to argue that the system is either efficient or even just, if it rewards poor work and disenfranchises good work. An application might get denied simply because the professor's student cohort has already received rewards in that year. However, if these rewards came from merit, then this is precisely the reason it should be accepted!

A second issue is institutional and governmental regulations for the use of funds. Institutions often limit the budget for a trip, having the researcher to spend the rest from his or her pocket. Another example, government and institute funds may only be used for Air India flight tickets. If you wish to travel another airline, you actually need to submit an application citing your reasons for not traveling Air India. An IAS officer will evaluate your application. This becomes another headache, both for the inconvenience of having to travel a suboptimal route on Air India, or for running around to get the necessary approvals to use an alternative airline. Our academics have better and more productive ways to spend their time. We should be supporting their efforts, not throwing ever more obstacles in their way.

What is the solution? Doubling the amount of funding and removing the approval processes? As if it were not impossible anyway, such an approach would be even more counterproductive. Funding should be driven by merit and performance. Faculty who are doing more work and better work should be rewarded with more travel funding while underperformers should get less. Distributing funds evenly penalizes the high performers and benefits the poor ones. In principle, there should be no barriers to travel for productive and passionate researchers. Our systems should support them enthusiastically.

In the United States, travel funding is largely determined by the faculty member himself or herself. The money comes from the funds he or she is able to raise for his or her projects. When writing the grant proposal for his or her research, the professor itemizes the sum that he or she expects he or she will need for the project. The more projects the professor has, the more grants she has: therefore, the more funds for travel. As an indicator of performance, the ability to raise funds might not be ideal. Some people are better at selling their ideas than others. However, research grants are mostly approved on the quality of the idea behind the proposal and also on the past performance and success of the researcher. They are "good enough" for giving away travel funds.

Travel funds encompass more than just the travel to a conference to present a paper. There are other uses also. One needs to travel to meet collaborators, to form new collaborations, and to deliver talks about one's work: all with the purpose of furthering the project's goals. Travel funds are also meant to provide for the travel of others: to bring collaborators to your own lab or to deliver a talk to you and your students on some aspect pertaining to your work. In India, there are no institutional mechanisms for such things internationally. Such funding is available for collaboration within India only. Generally, the speakers from international schools who appear at IITs are Indians on a personal visit to their home country.[5] A faculty member cannot fund the international travel of a colleague from any of the funds he or she might have. Our brain and cognitive science professor who is developing next-generation prosthetics would unfortunately not be able to invite Professor Robert Gaunt from the University of Pittsburgh. He or she also would not be able to travel to his university to get his input and advice. He or she would labor on as a "tortoise in a well."

Administrative Staff: The Need for More...

Faculty requires administrative support for such things as travel bookings, managing appointments, purchasing, financial accounts, reimbursements, and much else. Administrative staff performs many tasks on campus including campus upkeep, library management, help with alumni affairs, careers office, and so on.

In top US universities, each faculty member has an administrative assistant or shares an assistant with one or two others. These assistants manage the day-to-day work of the office. At the department level, there are additional staff to oversee various programs and departmental affairs. In contrast, Indian faculty members typically have no personal staff. Instead, the department has a pool of staff that the faculty may approach for help. For example, in a typical department at IIT Mumbai, about 7–8 departmental staff assist around 40 faculty members.[6] Our faculty rates availability of sufficient and

Table 5.1: Faculty, Student, and Staff Numbers for Different Universities

	IISc	IIT Madras	MIT	Stanford University	UC, Berkeley
Faculty	513	548	1,036	2,153	2,258
Students	3,743	8,961	11,376	16,122	38,204
Staff	833	588	4,071	12,000	9,020
Staff/Student (%)	22.3	6.6	35.8	74.4	23.6
Staff/faculty (%)	162.4	107.3	393.0	557.4	399.5
Students/ Faculty	7.3	16.4	11.0	7.5	16.9

Source: Annual reports, institutions' websites or communications offices.

capable staff the second lowest among 14 parameters of quality (below average, 2.5 on a 1–5 scale).

Table 5.1 compares the faculty, student, and staff numbers for some schools in India and the United States.

The number of staff members depend on some weighted function of both the number of students and number of faculty members. We can thus compare universities with similar student to faculty ratio. We find that Stanford University has three times more staff per faculty as IISC. Similarly, University of California, Berkeley has four times more staff per faculty. MIT again has a much higher staff-to-faculty ratio as compared to Indian universities.

These ratios suggest that our universities could use more staff. Staff helps faculty efficiency and productivity by freeing them from administrative responsibilities. In fact, given the level of bureaucracy and institutional regulation in India compared to the West, I would argue that staff-assisting faculty is more a priority for India.

Reading and Writing Papers

We have discussed the importance of physical equipment for research experiments. Even more important is the acquisition of knowledge to conduct real research. For this, one needs access to journals and papers. The access has improved exponentially for Indian institutions.

With the Internet revolution, papers suddenly became highly available to everyone in the world. Indian institutions now subscribe to various journals and services such as IEEE *Xplore*. Also, some papers can be accessed for free via services such as Google Scholar and Research Gate, on author homepages, and also by e-mailing authors directly to request a file. Institutions in the United States enjoy additional and more sophisticated services such as interlibrary loan programs that allow you to request books and papers that the home library does not carry. Another great service is that one can borrow an unlimited number of books. Indian institutions have fair access to literature. They may be a little poorer in access to technical literature, but this is not a limiting factor today.

Another place where we need actual help is in paper writing. The vast majority of technical literature is in English, which is not our native tongue. Indian schools and colleges do not stress writing skills. Our academics, including me, cannot write English with the same quality and ease as that of native-speaking academics.[7] In fact, Indian papers often get rejected on the basis of the writing quality. Worse, the subpar writing sometimes creates a bias in the evaluation of the technical content.

Our researchers would be helped tremendously by professional copywriting and editing services. Today, I pretty much use an international editor for all my writings meant for global consumption. If our institutions sponsored such services, our researchers could save much time and get published more often. They could better communicate with the global audience, thereby creating a larger impact with their work. Undoubtedly, it would help them garner more citations also. This small intervention would have an outsized impact.

6

Research Environment: Connecting, Collaborating, and Competing

A scientist is like a painter. Michelangelo became a great artist, because he had been given a wall to paint. My wall was given to me by the United States.
—Riardo Giacconi, a Nobel Laureate in Physics

Scientists prosper in the right environment, just like any other professional. But a scientist's job is not like that of a typical nine-to-fiver. Scientists busy themselves with trying to create new knowledge. Creating the right culture and incentives in which such work can thrive is more complicated than the usual.

Typically, a company must provide for three broad areas in order to help their employees succeed. First, the employees should have incentives to perform well. Second, they should have the inputs they need to do their jobs. And third, the organizational culture must be conducive to success.

Incentives include such things as progressive salary increments for improved performance, and promotion to higher responsibilities including management, bonuses, and lifestyle perks such as company vehicles and extra vacation (similarly, there should be clear disincentives for poor performance, such as the withholding of raises and perks, and ultimately termination). An employee's performance can typically be measured in terms of quality of work, number of projects completed, timeliness of delivery, client satisfaction, and employee effort such as number of hours worked and willingness to help at critical times.

Inputs include the materials that the employee needs to carry out the tasks of his or her job—good infrastructure, computers, seating space, etc. More importantly, it includes proper training and ongoing skills enhancement. At least as important is culture. A helpful boss and congenial peers are essential for success. Reasonable salary and up-to-date software do not compensate for a hostile environment. A discouraged employee will never perform at peak ability and will spend much of her time dreaming of a better situation and looking for another job. If each of these components is in place, the organization has a good chance for success.

Some of these factors carry over to the university and the role of researchers. But there are enough dissimilarities to make the two incomparable. For example, scientists do not typically have bosses. They are more like partners in a firm, such as you find in consulting, law, and accounting. In most respects, they are their own bosses. No one monitors how much time they spend in their offices or how they spend their time. A true researcher, a dedicated researcher, will often work 14–17 hours per day, but if asked to report their time they would probably resign. They like their independence and chart their own course, not without guidance and mentoring from their more experienced peers. The long summer and winter breaks add in time for independent work. Their performance, especially in the short term, cannot easily be measured and is typically determined by peer review only (e.g., the opinions of their colleagues, measured over the course of years or decades).

They cannot be fired once hired or after tenure, typically 7 to 10 years after recruitment. The promotion ladder is generally short and has only three designations: assistant, associate, and full professor. Most good researchers reach the highest level within the first 20 years of their career. Their salaries stagnate. They may advance to other positions such as dean, department head, or university president. Although recognized as higher positions, these are primarily administrative, and most researchers avoid them because the responsibilities would take them away from their work.

In these unfamiliar conditions, what can we do to encourage researchers to perform well? What are their incentives? If they cannot be fired, how do we discourage underperformance? What drives a researcher to perform well and achieve breakthrough results that benefit the economy and the public good? How does innovation happen?

We will examine these issues in this chapter and, in the words of Riardo Giacconi, try to figure out how we can provide a wall so beautiful that Michelangelo himself would want to paint on it. Considering the university research environment, we will examine the nature of the researcher's interaction with four distinct groups: the administration, peers, students, and the international community. The interaction with administration concerns formal institutional rules of incentives and disincentives, promotion, teaching hours, and so on. Interactions with peers, students, and the international community are also influenced by institutional policies, but other factors in the ecosystem are even more consequential. These relationships are determined by such things as collaboration, competition, and even sheer numbers of like-minded people.

Each of these relationships is necessary for a healthy research ecosystem. One might think that a scientist sits alone in his small room with outgrown hair, shabby clothes, avoiding people and performing some strange undiscernible magic of his own. The reality could not be more different.

Interactions with the Administration: "The Carrot and the Stick"

Our best institutions—IITs/IISc—do not "require" that the faculty perform research. You can continue in your position even if you produce little research, mediocre research, or no research at all. The institute will not ask you to leave. Your salary will continue to grow, although possibly at a little slower rate. The Ministry of Human

Resource Development (MHRD) will continue to fund your PhD students.[1] You will receive your equal share of travel money. The only possible repercussion is that you might not get promoted. You might stay as assistant professor for life.

This does not mean that all faculty members in the system are ineffective. It simply means that the system does not push them to perform. A professor might nevertheless have strong personal motivations and ambitions that drive him or her to achieve. But, consider the numbers. Analysis of one of the IITs reveals that 40 percent of faculty members contributed 10–15 percent of the institution's research output. By contrast, 25 percent contributed more than half. The top 25 percent are workaholics, who work more than 12–14 hours a day.

Low performers do not incur penalties for their low performance. Do high performers receive rewards? A little. They can expect quicker promotions and a little quicker salary growth. They might get a chair position, which sometimes comes with additional salary. Given that they probably write good grant proposals, they might get additional money for students and for buying equipment, but this is a result of their own efforts, not institutional favor. They will not gain extra influence over the department or institution. Influence is reserved for the senior professors, appointed by age and seniority. Further, it is more available to those who know how to navigate the bureaucracy and institutional politics.

In essence, a socialist attitude pervades our institutions. We distribute rewards evenly, with no disincentives for slack. Underperformers can live a comfortable life—an incentive for the less motivated to join our ranks. Good performers enjoy some favor and really need to fight their way to get more.

One might point to the US system and ask how is it different? There also, a professor cannot be removed after the first 10 years or so of employment. Then, what is all the hullabaloo about?

The US system has some strong distinguishing features that make it work. The first is something called "tenure." In the tenure system, all first-time faculty join the institution in a nonpermanent (i.e., nontenured) status. The person will be offered tenure based on his or her performance in from the first 5–10 years, depending on performance. If he or she accepts, then he or she has "tenure" and cannot be removed (barring some egregious violation). Her work and ambitions meet with approval, and he or she is granted the freedom and independence to pursue his or her research path.

The tenure process at top research institutions is highly competitive. At these schools, anywhere from 30–60 percent of junior members fail to win tenure. From 1991–2004, MIT denied tenure to 53 percent of its aspiring faculty. Realize that these faculty members entered the process through an already-competitive selection process. There were already the best for the best. In his book *The University*, Professor Henry Rosovsky describes the criteria for gaining tenure at Harvard: "who is the best person *in the world* fitting the job description? If the best person is—in the opinion of departmental colleagues—one of our own junior members, he or she will be nominated for promotion." The nontenured faculty has to compete with everyone in the world to win tenure.

An article in *MIT Tech* explains the criteria for tenure at MIT this way: "you are either the top investigator in your field, or one of the very tiny handful of top investigators in your field, in the world."[2] Another professor at MIT takes it further: the criteria is "that a person will improve the reputation of the institution." The process of tenure is quite complex. Other than a detailed analysis of all the work of the person, it also depends on recommendations from professors within and outside the institution. As Rosovsky explains in his book, one of the things they ask external reviewers is who are the top five researchers in the field? Then they note if the candidate's name comes in and at what rank. The candidate then undergoes two to four levels of review by the departmental faculty, followed by another review by the academic council, which includes the president of the university. Finally, the university board gives its seal.

I have half a dozen friends in the United States who are now on a tenure track or who have been awarded tenure recently. These folks work crazily hard and are under enormous pressure. Who is a leader—the subjectivity makes it all the more tough! I consider that these folks are doing a job very similar to mine—that of an entrepreneur. They have to conceive new ideas that may prove pathbreaking. Then they must pursue them: raise money, set up a lab, and attract and hire good students. Concurrently, they must teach a class, possibly two (teaching a class for the first time can be quite unnerving at a world-class university, especially when you know that the students are very smart). Preparation for the course and for each class takes many hours.

Additionally, the aspiring faculty will have departmental duties such as a course curriculum committee or some industry liaison work (when you are pursuing tenure, you cannot say no to any request for your participation!) And despite your best efforts, you may still fail. Your course might receive poor reviews, the grant evaluation committee might reject your proposals, or your experiments might not deliver satisfactory results. So despite your long years of effort, the department might deny you tenure. Just as an entrepreneur pours her money, hopes, and dreams into an enterprise, success may simply slip through his or her hands.

However grueling the tenure process, it prepares a young PhD to understand the process of research, get soaked in the research culture, and become a world-class researcher. He or she produces some of his or her greatest work in his or her early years. Many faculty assert that their pretenure years were their most productive and exciting time.

In this process, the institution gets a chance to evaluate and weed out who aren't good enough. For the selected, tenure once bestowed cannot be taken away. However by this time the person has already produced a formidable cache of research and they have fallen into a "habit" of high performance, and they are highly unlikely to stop.

Unlike in India, faculty members in the United States have disincentives for nonperformance. For example, top schools in the United States have a distinct salary system whereby the faculty receive payment for only nine months. The professor himself or herself must make up the rest with grants—mostly during the summer break. He or she will write grant proposals for summer research work, itemizing summer salary as a portion of the grant. If he or she does not propose any summer research and prefers to take it easy, then he or she forgoes that 25 percent of his or her payment. Furthermore, grant proposals account for all the professor's travel money and student research money. If he or she does not pursue them or win approval, then he or she will simply have no funds for these things. All resources tie into one's ability to raise funds. If the department begins to notice a pattern of underperformance, they could take other steps such as assign him or her a smaller or poorly located office, or even take away his or her lab space if they feel it is not being utilized well. However, the greatest disincentive will always be the loss of reputation among peers. We will discuss this in more detail later.

On the other hand, university system empowers successful and high performing researchers with professional rewards. They give them better access to facilities and give them a larger say in the department's decision-making. Furthermore, high performers can negotiate much larger salaries for themselves. Like corporate super-stars, they can even leverage offers from various institutions (by contrast, Indian academic salaries are mostly fixed, with little vari-ance). Even public schools, where salaries of all faculty members is public information, indulge in negotiations to retain their best. They run the headache of paying for performance yet consider parity, like any corporation. Further, the department may provide more space, supplementary research funds, beyond what the faculty person could raise on his or her own.

US universities provide great professional rewards as well. They open up facilities for them, such more and better lab space, a larger and first choice on choosing PhD students, providing supplemental funds and so on. To amplify the success of their superstars, they get more faculty in the same area and provide the superstar faculty greater say in such decision-making. The researcher gets more col-leagues to discuss, develop, and pursue new ideas. They may also get some breaks in teaching responsibilities. Finally, the department and university will put a spotlight on the researcher's work in all their marketing material and nominate him or her for awards and to highly reputed bodies like the National Academy of Engineers. The department may extend chair professorships to them and eventually provide them the highest honor in the school, that of a university professor.[3]

Chinese universities also provide direct benefits for publishing papers. In 1999, Shanghai Jiao Tong University (SJTU) began awarding monetary prizes to researchers who successfully publish papers in top journals and conferences. Part of the award is apportioned to the researcher's projects but the other part is for personal use. These awards range from $120–$150 for an average paper to $1,480 for a highly influential paper. Such an award policy is somewhat controversial among academics; nevertheless, it demonstrates the universities' commitment to encouraging productivity and high achievement. In conclusions, the US university system "pushes" their faculty to perform. The incentives and disincentives are not quite like those of a typical job in the private sector. But they do exist and serve their purpose. In India, the professors must push the system to receive the resources, benefits, and environment.

Mind Space and Time for Research
Social Infrastructure

Researchers function best in a hassle-free environment. If one is going to ponder great questions, one needs space and quiet to think. The cares of modern life—water issues, electrical brownouts and blackouts, traffic problems, school, and childcare issues—all make demands on one's time and mindshare.

Our campuses can do a great service by providing a functional hassle-free physical environment. Already, they do moderately well, offering benefits not typically found on Western campuses: subsidized faculty housing on campus. Other standard facilities include medical care and banking services. However, only a few universities offer extra amenities such as child day care, primary/secondary schools, and gymnasiums. Moreover, existing facilities are often plagued by undercapacity, nonavailability, delays, and poor maintenance. The remote location of many universities means that it is not easy for the spouse of the professor to find a suitable job. Furthermore, there are often no good schools in the vicinity for the faculty's children. School admission offer added hassles—limited supply and corruption.

Our universities cannot provide for everything, and we should not expect them to. But much can be done to assist basic day-to-day needs. We should have a modest well-functioning system. A little effort by our institutions would send a message that they care and that they value the researchers' time and work. We shouldn't leave our researchers to fend for themselves. A little effort could go a long way.

Teaching Responsibilities

Teaching is an integral part of a faculty's duties. Teaching creates the next generation of researchers and professionals. Furthermore, a vibrant course can add to the professor's thought processes and deepen his or her understanding of his or her subject. Teaching is also a great way to learn a new subject. A faculty at MIT remarked that he wanted to learn a new area of his field, so he decided to teach a course on it. Being responsible for a course's content will quickly bring you up to speed!

Despite all the benefits of teaching, we must acknowledge that it also takes time from research. Teaching and research must find a proper balance. The teaching load for research faculty is typically one course per semester. This holds true anywhere in the world. Our faculty may require to do a second lab class or coteach another class. Teaching load is more for smaller departments and those with challenges of filling positions.[4] The inability to fill positions is a constant challenge at many schools. This puts additional load on current faculty members. Additionally, US professors may sometimes have flexibility to skip a semester and make up for it later.

Although teaching commitments between India and the West may look similar, a larger difference is created by the lack of good teaching assistants in Indian institutions. PhD students take on teaching responsibilities with their professors, in addition to research. They help in preparing material, answering student queries, and grading homework and exams. In many universities, PhD students are required to teach at least one such course to earn their degrees. This is a good learning experience for the PhD student and is a great help to the department.

Our top institutions have PhD teaching assistants also. However, they generally do not do the heavy lifting and support the faculty in the same way as their counterparts in the United States. One faculty member from an IIT remarked to me that his undergraduate students are more reliable and helpful than his teaching assistants. In this school, master's degree students of technology had compulsory teaching duties as dictated by the MHRD as a condition of their financial support. However, the teaching duties of PhD students are decided by the students' advisors. They usually end up with no teaching responsibilities at all. One MIT-trained faculty at another IIT remarked to me that the teaching assistants in India were much lower quality than one would expect to find at MIT. This lack of support in teaching by the PhD cohort is a function of the quality of our PhD students, but also of the trust of the faculty. A better teaching assistant culture could help create a better research environment for the faculty and for the institution. We will discuss some of these dynamics between faculty and PhD students later.

Interactions with Peers
Collaboration

Scientific collaboration is a social process and probably there are as many reasons for researchers to collaborate as there are reasons for people to collaborate.
—J. S. Katz, 1993

Speaking is not just a means of communication, but a means of "intelligence." This is what I learned when I first browsed through the websites of Marvin Minsky and his student, Patrick Winston, both pioneering AI and MIT professors. They gave a simple example to illustrate the point. Oftentimes when we plan to ask a question to someone, we know the answer as soon as we articulate the question. It happens to me all the time. It is a funny situation, where I utter the question and then immediately follow it up with "I got it." The process of pushing our unorganized thoughts through structured language helps to clarify our thoughts and adds value to them. One may sit alone and think for hours about a problem without making any progress toward an answer. But we find the answer in a jiffy as soon as we discuss and try to explain our thoughts and questions to another person. But for this to happen, we need to be speaking to someone whom we are fairly confident of and who understands our subject, has an interest in it, and is ready to listen. My advisor at MIT was one such person who got the best out of me in these discussions—and she added formidable value to it herself.

This small example illustrates the importance of communication in the life of a researcher. Communication does more than clarify one's own thoughts. Communication with great colleagues helps one to see a larger picture by exposure to complementary or competing techniques, their experiences, and results. The feedback of knowledgeable colleagues helps the researcher to see his or her problems in a new light and to get unstuck. The enlightenment may come in a moment through colleagues' experience or "tacit" knowledge.

Tacit knowledge is that which is unexpressed, not explicitly known to the doer but just in his or her head. It is not easily covered in scientific publications or documentation; however, its role cannot be underestimated. We can recognize its value more often in experimental work, where building a robotic arm or developing a microfluidic chip requires some unwritten skill that comes with experience. We also see it in the development of new algorithms, especially if they are heuristics or recipes to solve problems such as how to automatically classify images. Deep neural networks, one of the most advanced AI algorithms, is a heuristic. In my own experience, when colleagues say that genetic algorithms did not work for their problem, I often wondered they were yet to figure out how to make them work. Tacit knowledge is what makes replicating results so difficult, which constitute a key skill that a PhD student must acquire. Tacit knowledge is exchanged in informal discussions, mentorship, collaborations, or through direct partnership with people who have this knowledge.

Today's interconnectivity does not replace geographic proximity. Discussions and the sharing of tacit knowledge does not easily happen remotely. Phone calls and e-mails are good means to communication, but they do not constitute "means of intelligence" in the way that personal face-to-face conversations do. Anyway, one of the most difficult thing about a conversation is initiating it. The most meaningful conversations and productive conversations begin informally. One cannot initiate an informal discussion via e-mail or phone. Once initiated, collaborations make more progress during a single personal meeting than in several exchanges of e-mails and phone calls.

There is evidence for this beyond the anecdotal. J. S. Katz, in 1994, analyzed data from Australia, the United Kingdom, and Canada to study how geography influences research collaboration. He found that research collaboration decreases exponentially with the distance separating researchers. Frenken et al. in 2007 studied data from the Netherlands.[5] They found that collaboration among different kinds of institutions—universities, government institutes, and private businesses—is most strongly influenced by geographic proximity. Institutions that were locally situated were more collaborative and generated more economic value.

Easy and informal access to intelligent research colleagues greatly enhances the productivity of researchers. Communication leads to more cross-fertilization of ideas, citations, and research impact. Furthermore, communication facilitates collaboration. Most new collaborations begin through random casual conversations. Ten researchers working in close proximity will be far more productive than the same 10 researchers working in isolation. We need a critical mass of researchers with easier communication channels to make greater progress. They can multiply each other's successes.

Ecosystem Effects

A critical mass of researchers situated in a geographic region has practical implications for mutual assistance. Proximity creates a feedback loop that leads to greater resources for research and larger impact. Institutions can invest in large, expensive equipment and infrastructure to be utilized across many researchers and many research projects. Sophisticated and expensive equipment will produce a higher rate of return through the combined impact of several research projects. It also creates substantial demand for research hardware, manufacturing skills and process. Tangible financial returns lead to the growth of local vendor networks.

A critical mass also facilitates the availability of research manpower. When a professor needs to hire a postdoc or add another faculty in his or her area, he or she has more to choose from because the PhD cohort has already gained much experience through the wider network of professors. Trained personnel not only facilitate the speed of research projects but also assist the transfer of tacit knowledge and the cross-fertilization of ideas. For this reason, some US universities do not recruit their own PhD students to tenure track positions immediately, but require them to spend time in a different institution with a different set of colleagues.

Eventually many of these skilled PhD students go to work in the private sector. The critical mass carry their ideas into the private marketplace, pushing industries to adapt new methods. They also become entrepreneurs. The many entrepreneurs in an area share knowledge, learn from each other's failure, and some eventually succeed. They feedback resources for researchers such as industrial infrastructure for research, employment opportunities for PhD students, research collaboration, and sponsorship. These effects show only when there is critical mass. A few trained PhDs get lost as noise in the routine of industry life. Only a few become entrepreneurs, and they suffer from lack of like-minded people, trained manpower, an ecosystem, and have a high failure rate.

We see great examples of this around Stanford and the Boston-area universities. Companies founded by Stanford people in Silicon

Valley now generate $261.2 billion in revenue—55 percent of the total revenue generated by the "Silicon Valley 150" companies.[6] A 1997 Boston Bank report found that Massachusetts had 1,065 MIT-related companies, constituting 5 percent of total state employment and 10 percent of the state's economic base.[7] MIT-related firms accounted for 33 percent of software sales in the state and 25 percent of all manufacturing sales. To add, there is considerable research evidence concerning the value of geographical clustering of research and innovation.[8]

This critical mass of researchers can potentially create large impact in their local geographic area. The law of large numbers and network effects makes progress, and even disruptions, most likely to occur. Those away from the ecosystem are in some way "off the center of gravity," as one Indian researcher remarked. They are "followers"—labor away on their own little projects, occasionally connecting with the true ecosystem and the major players to find a yardstick for the progress they have made and to collect a new set of problems to occupy their time for the immediate future. Where does India stand in the global research ecosystem?

In the Shadows

In our survey, researchers reported that "having a critical mass of researchers in one's field" is among the poorest 5 of 14 parameters of the Indian research environment; 74 percent faculty members rate it average or below average. As noted earlier, India has a much lower number of high performing researchers compared to the United States and China. As we slice and dice this already-low number by fields and subfields, the researchers per field are reduced to two-digit numbers. Some emerging fields such as brain and cognitive science have among the fewest researchers in India.

Let us consider AI, where we are only slightly better. In terms of output, Tsinghua University in China produces more papers in data science than all the Indian institutions put together, each year. If we look at papers from Neural Information Processing Systems (NIPS), the leading machine learning conference, the United States has contributed 1,179 papers, China 110, and India 38 (2012–2015). Similarly, in Knowledge Discovery and Data Mining (KDD), the United States has contributed 497 papers, China 116, and India 28. Our paper count is in lower two-digits. Clearly, we have no critical mass and are off the center of gravity. The research and innovation ecosystem is owned by the United States and China.[9]

They define the research direction, the research questions, and they make the most important progress. Our small body of researchers make small collective progress. There is insufficient opportunity to discuss, collaborate, and share knowledge and resources. Institutional relations, local workshops, and conferences that help spur collaboration are also miniscule.[10] The variety of research directions and techniques are limited, leading to little cross-fertilization of ideas. The West script the next breakthrough in the field. We learn from them and then again make our small improvements, until the next innovation. With these small improvements, some of our researchers become superstars in our small community. There isn't as much local competition for them.

It is harder to convince the institutions or the government for expensive infrastructure without a critical mass to use them. But without these resources and without much obvious prospects that they are forthcoming, it becomes more difficult to attract the right people to research careers. This also limits opportunities for knowledge transfer to the industry and creation of economic value. Critical mass in research is a necessary condition, if not sufficient one for industry to be innovative. Certainly, this one ingredient is missing.

We live in the shadows of the global research ecosystem.

A Culture of Excellence

Peers interact through competition, not just collaboration. At world-class research institutions, competition among faculty can be fierce. Professors seek to outcompete each other in top journal publications, citations, grant acquisitions, start-ups, and awards such as the MacArthur Genius Award, and even the Nobel Prize. Everyone is a high achiever, each in his or her own way. Scientific discovery and advancement remain always the primary motivations for scientifically minded people. However, at top institutions such as MIT, if you do not achieve, you will fall from the level of "normal" and risk losing the respect of your peers. To achieve is the normal. Everyone puts in the hours, the hard work, handle multiple responsibilities, and take their work very seriously. Even as this is a norm within MIT, these faculty members features in the top 5 percent outside. Some go beyond the normal within the institute by becoming a university professors, win a Nobel Prize or a Fields Medal.

In the world's top research universities, excellence is a distinguishing feature of the culture—a way of life. This culture is cherished and valued, and everyone constantly strives to up the bar. Such culture maintains all aspects of life and work for a person. Most discussions you have during downtime at the institution—during lunch or coffee—revolve around science or some other intellectual topic. It is a nerdy community for whom "fun" is always comprised of scientific matters. A constant feature is the "can do" positive attitude and a constant belief in humankind's potential to create new knowledge.

India's top institutions feature a very different equilibrium. Here, most faculty members satisfy themselves with fulfilling their teaching responsibilities, publishing a few papers and managing their administrative responsibilities. A large proportion of faculty at our institutions are happy doing a 9–6 job and seek a work-life balance that tilts toward relaxation. Your colleagues do not disrespect you if your research is not world-class. A few faculty do achieve highly despite the system, but they are outliers. They cannot move the system to a new equilibrium of excellence. Excellence and mediocrity are each stable points of equilibrium—our institutions have gravitated to the mediocrity.

Why we have mediocrity as an academic cultural trait is anyone's guess. One may argue that it is just due to a lack of incentives for performance. However, I am not convinced. Even the presence of incentives might not improve performance. If none gets the incentive, they all still remain equal. Excellence has more to do with individual and group motivations and norms, not just incentives.[11]

I believe part of the answer lies with what is recognized and celebrated by the institution. One professor at a US institution told me how each semester the dean calls a meeting to discuss how much money was raised by various departments, which faculty received awards, and whose published papers are noteworthy. She was impressed by how the dean compared the various departments—not for shame, but in a positive way that instilled admiration and encouraged everyone to perform highly. This person also worked in a top Indian institution and seldom found such encouragement there. The answer lies here.

Partnership Between Faculty and PhD Students

Research faculty works with PhD students to produce research results. For a productive relationship, a research professor should have a great working relationship with his or her students. In a healthy environment, PhD students are considered worthy colleagues, who contribute productively to ideas, who have good ideas of their own, and with whom the faculty member can explore and probe his or her own thoughts. They are not just subordinates who have to obey the professors' orders, like hired labor. They work on projects together, examining problems and deciding on solutions. Although they work under supervision and mentorship, the students enjoy a measure of independence. As members of the upcoming faculty generation, the students compare (and compete) in intellectual prowess with the professor. They become "equal partners in research," as described at MIT. This respectful relationship forms the bedrock of the success of the US research ecosystem. This is in contrast with the traditional German model of "apprenticeship," where a PhD student is a subordinate to the advisor.

This is not to overstate the equality of the relationship. In fact, a major part of a PhD student's time is spent on heavy lifting: reading papers, solving equations, building robots, writing code, designing circuits, counseling undergraduates, and many other things that a professor would like to pass off in order to stay focused on the central thrust of his or her research. The student is the workhorse of a research project and must execute well. But the student's intellectual ability is the key to his or her ability to execute these tasks. He or she must take many micro-decisions correctly, which will have significant repercussions for the endeavor if he or she decides incorrectly. In fact, the professor chooses his or her students precisely on their confidence in their ability to make the right decisions at critical times. In my work, I appreciate research engineers who can inform my research ideas and who are also great in execution. I encourage them to surprise me by going beyond how I asked them to approach a problem.

Over the course of his or her PhD mentorships, a research professor sometimes finds a "star student" who excels on all parameters. He or she becomes the professor's right hand and begins to assist with the professor's entire agenda: revising papers, mentoring fellow students, helping in conference/workshop organization, and facilitating collaborations. With such an assistant, a professor typically performs his or her best work. Star students enhance the productivity of the entire research group, even allowing the faculty to take on additional students. All faculty dream of finding such students. The two working in tandem create a virtuous circle with each assisting and feeding the other's progress and growing the body of new knowledge together.

I recall Kumara Sastry to be one such student for David Goldberg, a pioneer in genetic algorithms. Kumara spent nine years at University of Illinois Urbana-Champaign (UIUC) with Goldberg, virtually taking the lead in running his group. I also had my own star student in Shashank Srikant, a passionate engineer whom I hired from NIT Kurukshetra. Shashank turned down an offer from Microsoft to join us. He stayed five years, taking up day-to-day mentoring for the research group members, helping me finalize four papers, organizing workshops, and running ml-india.org and datascienceindia.org. He still remains an active collaborator.

In India, this relationship needs to be nurtured. As discussed in Chapter 3, our institutions do not attract the best undergraduates to PhD programs. Most PhD students are not high achieving, and they lack sufficient training in research processes from their undergraduate days. Only 35 percent of faculty members rate the quality of their PhD students as either good or excellent. Faculty spends a considerable amount of time bringing their students up to speed on basic research processes. Time spent per student is high. The necessity for constant oversight slows the pace of research projects and negatively affects the quality of results. The time-intensive mentoring also limits the number of students that any single professor can take on. Star students rarely appear and the faculty doesn't get a chance to leapfrog their productivity.

We need to create a better culture in our institutions. Students often remark that faculty and institutions act as if they are doing a favor by having them. This undermines their confidence and diminishes their capacity for productivity. The advisor adapts a manager–reportee style than a partner relationship even with the more capable students.[12] The faculty and the institution must address these issues with constructive steps. We need to develop a culture of mutual respect and partnership. One PhD student noted in our survey:

> Aside from TIFR, there are only a few institutes that provide time for free thinking, as well as peer pressure. In many Indian institutes, PhD students feel disrespected by their advisors. Indian PhDs need to be respected, in India first and should be considered on par with their counterparts in similarly-ranked institutes abroad.

If given the chance, 56 percent of PhD students in India would prefer to go abroad to study, and 36 percent of them would not recommend their friends to follow their path. They do not hold their research programs and advisers in the highest esteem. Their idols are rather in the West. Whereas Western institutions enjoy a positive feedback loop from professor to student, the comparable relationship in India contributes to a low-efficiency culture. This situation further diminishes the perception of a PhD/research career, further discouraging talented students from joining, and further ensuring that research ranks will be replenished with more mediocrity.

Interaction With Global Community

आख़िर जब मैं आसमान में, उड़ी दूर तक पंख पसार, तभी समझ में मेरी आया, बहुत बड़ा है यह संसार
Translated as "When I spread my wings to fly in the open sky, I realized then, how large the world is."

—Final lines of the poem "इतना सा ही है संसार" (Small World) by Nirankar Dev "Sevak"

Scientific research is a global endeavor. One has to solve a problem or answer a question, which no one in the world has done. Researchers create new knowledge for the world with relevance beyond borders. Research uses and creates knowledge, which has relevance beyond borders. Their work is benchmarked on a global scale than an institutional or a national scale. This is in contrast to other professions and organizations where competition is local or, at best, national.

Researchers must engage with the global community to actively participate in projects, exchange knowledge, create competitive new knowledge, and to spread the impact of their research. Given that we have no critical mass of researchers, the need for global interaction is even more important. Let us consider some benefits of international travel, which are as follows:

1. *Hear and learn:* Interacting with people personally helps one know the latest developments in the field, the nuances of different approaches to a problem or process, and the general

discourse in the community. Conferences and talks allow one to hear new research results directly from the authors, sometimes even before publication. One gets to know the community impression of a new work, ask their doubts, and discuss ideas of new work. The opportunities afforded by direct communication with authors over the course of a two- or three-days conference would take six months to realize if pursuing a correspondence by remote. Furthermore, taking time to listen to what others are doing can give you a fresh perspective on your own work, by taking your mind off of it and forcing you to think about the field from a different angle. It creates a balance between exploration and one's own work progress.

As discussed earlier, face-to-face meetings facilitate the transfer of tacit knowledge. Technical literature tells a linear story of "success." A write-up does not typically describe the challenges, failures, implementation details, and hacks of the process. Nor does it discuss how the technique may be generalized or what its real-world relevance might be. A colleague at MIT had coined a phrase: "literature lies!" The research that does not get published is at least as important as the research that makes it into the papers. To get at these important matters, we must speak directly to the authors and discuss with the community.

Conferences help researchers develop a "big picture" understanding of the field. Group discussions decide which problems can be considered solved, which have simply become obsolete, and which new problems are rising up. Through discussion, we decide the kinds of problems that are most interesting and worthy of pursuit, what are the most promising approaches and techniques, and what are the big challenges ahead. In two–three days at conferences and workshops, one senses the pulse of the research community and their group wisdom. One doesn't need to follow the community, but get inspired and create a defense for his or her research ideas in the community. One must consider that the field of AI was formally born at the Dartmouth Conference in 1956: a real watershed moment in the history of computer science. At this conference, AI got its name, its mission, and also its first set of successful researchers.

2. *Tell and sell:* In addition to learning about other's work, conferences and talks at universities allow one to inform the community about one's own work. In graduate school, we used to joke if one person other than you and your advisor read your journal paper it is a great achievement! By contrast, in a talk, you get an interested audience for 15 minutes to an hour to hear the details of your work. This is supplemented during downtime when you have individual informal discussions and meetings with fellow researchers.

One needs to sell one's work. It leads to three things: (a) generates value and impact; (b) gets the researcher credit; and (c) helps develop the research field. Conferences and talks inspire others to use one's work, develop it further, or apply it on new problems. The researcher's work garners credit in this way and contributes to the development of the field. As the work accrues recognition, the researcher's resume grows in stature. They get nominated on editorial and program committees, are invited for keynotes, and it becomes easier to win grants. The low citation rate for Indian researchers is not just attributable to low quality but also to our poor efforts to promote our work.

This process of doing good research, talking about it, inspiring other work, and collaborating is what helps real useful new knowledge to develop. It takes years for research to create value, and "network effect" is an essential ingredient. If you have a revolutionary new idea, have some evidence for it, and want the community to coalesce around it, you need to orchestrate a plan of talks, visits, workshops, and conference presentations. You cannot just push it through your research group and hope that it catches on.

Researchers present their research through talks at universities and organizations, other than presenting papers at conferences. I have presented my research work at several universities and research labs in India, the United States, Australia, China, and Hong Kong. Many of the people that I met at these places now regularly cite my work. Further, it has led to international collaborations, great intellectual friendships, and organization of workshops at conferences.

Some of this travel was paid by the host institution. For some there was partial support, and for some there was none. We should make it possible need to be proactive and create funding for our researchers to advertise what they are doing. A small investment in this way can bring great returns in popularity and consequently, greater availability of external funding.

3. *Collaborate and recruit:* As alluded to earlier, face-to-face meetings facilitate communication and collaboration in a way that e-mails and telephone cannot. In 2013, I began a collaboration with Professor Steven Stemler of Wesleyan University. Over three years, we exchanged countless e-mails, calls, and several paper drafts. However, all of these exchanges combined were not as productive as either of the two personal meetings that we held in Boston and Connecticut. In the first, we laid out the plan of the argument, and in the second, we decided on the proper interpretation of the results. The paper came out in October 2016 and was written up by *The Wall Street Journal.* Other collaborations have shown less progress because my partners and I have not had the opportunity to catch up in person. In fact, I have come to see face-to-face meetings as a necessity. Such meetings require one or the other to travel. Again, travel requires cash. No cash, no travel, no meetings, no progress, no results.

Travel also facilitates recruiting. Conferences and talks give PhD students the opportunity to advertise themselves as well as their research. Through presentations and discussions, they may land a postdoc or faculty opening. Furthermore, the student's adviser can schmooze and talk up his or her students to other colleagues who are always on the lookout for new department recruits. Additionally, private companies are now regular attendees of top research conferences, particularly on data science and human–computer interaction (HCI). Amazon, Microsoft, and Facebook sponsor lunches, dinners, and other events at KDD, for example, with the purpose of identifying new talent. When so many bright young Indian people cite low job prospects as a reason for avoiding the research path, access to events such as this make a great persuasive case.

Our International Interactions

With all of these potential benefits, how well do our Indian institutions facilitate international interactions? According to our survey, not very well. PhD students rated international collaboration as second from the bottom among 11 listed factors of PhD quality. Only 16.8 percent of PhD students think the degree of international collaboration is above average. Table 6.1 reports international visit numbers for some of the top Indian research institutions.

Although global engagement exists, it is less than optimal. The percentage of faculty who make international visits ranges from 17 percent at IIT Madras, mechanical engineering to 75 percent at IIT Madras, computer science. However, the number of annual visits taken by each faculty on average is less than one. A well-performing faculty member should be expected to engage in international visits three times annually, or at least twice. Only around a fourth of PhD students do an international visit in a year. The number of visits should be at least one for every two years, or even once a year, of his or her student career. India's global engagement is half to one-third of what is typical for a healthy academic research environment. A remark by one student from a top institution sums it up well:

Table 6.1: International Visits by Students and Faculty at Top Indian Universities

	IIT Madras Computer Science	IIT Madras Mechanical Engineering	IIT Bombay	IISc
PhD students	79	302	2,500	2,490
Total faculty	28	64	584	513
Number of student international visits	21	37	553	NA
Number of faculty international visits	21	11	335	264
Student visits/ Students (%)	26.6	12.3	22.1	NA
Faculty visits/ Faculty (%)	75.0	17.2	57.4	51.5

Source: Annual reports or institutions' communications offices.

Notes:
1. IIT Madras data is from 2015–2016. IIT Bombay and IISc data is from 2014–2015. This is as per availability. The data for IIT Madras includes two representative departments.
2. IIT Bombay has not provided the exact number of PhD students for 2014–2015. Instead, the data includes the numbers from earlier years and projections for later years. The number reported here is an approximation.

Problem statements are not current enough (at best, India is about 6 months behind the USA). Not enough fellowships/scholarships (although this situation seems to improving) and very little travel funding. How are students supposed to meet up with other researchers and start collaborations if we're not even allowed to go to conferences? Very few opportunities to pursue research after getting a PhD.

How many speakers from foreign universities present at top Indian institutions? Table 6.2 provides some numbers, breaking out people of non-Indian origin. This is important, given that many of the Indian nationality speakers are merely presenting as an add-on

Table 6.2: Number of Talks by External Speakers by University

	IIT Madras (Computer Science)	IIT Madras (Mechanical Engineering)	IIT Bombay (Computer Science)	IISc (Whole Institute)	SJTU (Mechanical Engineering)
Total faculty	28	64	47	513	384
Number of talks by speakers of Indian universities	6	7	23	410	NA
Number of talks by speakers of foreign universities	10	14	24	460	300
Number of talks by speakers of non-Indian origin	0	11	12	NA	NA
Year	(2015–2016)	(2015–2016)	(2014–2015)	(2014–2015)	2012
Talks to faculty ratio (%)	35.7	21.9	51.1	89.7	78.1
Foreign origin faculty talks to faculty ratio (%)	0.0	17.2	25.5	NA	NA

Source: Annual reports or institutions' communications offices. IIT Madras data is from 2015–2016. IIT Bombay and IISc data is from 2014–2015. This is as per availability. The data for IIT Madras includes two departments. Talks at IIT Madras were listed as "Distinguished Talks."

while they are on their annual visit home to see family (i.e., they did not make a special trip in consideration of the importance of their work). Our researchers should have opportunity to hear the great work of non-Indians as well. The total number of talks ranges from 10–24 for a single department, annually. For speakers of non-Indian origin, this number goes down to 0–12. As an example, Computer Science and Artificial Intelligence Laboratory (CSAIL) at MIT hosted 42 talks from external researchers in April 2016. Realize that CSAIL is just one of four major labs in the Electrical Engineering and Computer Sciences (EECS) department, and there are more than a dozen such departments. On average, the university hosts 100–200 talks per month, campus-wide. MIT might be an unfair comparison, but even mechanical engineering department at SJTU hosts about 300 talks per year.[13] A fair metric is approximately one invited speaker per year per faculty member. The IITs achieve about a quarter of this target, if we consider speakers of non-Indian origin.

IISc has a better record of invited speakers than the IITs. IISc hosts approximately 460 talks annually by speakers from foreign universities or 89 percent of its faculty number. As India's flagship research university, it attracts a number of foreign delegations. However, it still pales beside MIT and SJTU.

There have been some recent initiatives to increase international interactions. For example, Machine Learning Special Interest Group (MLSIG) at IISc and Data Analytics & Intelligence Research DAIR at IIT Delhi have been adding guest speakers to augment their machine-learning programs. DAIR organized 22 talks in 2016, of which 14 speakers came from universities outside India, and 8 were non-Indian.

Now let us consider international collaborations. Indian and Chinese institutes perform the least number of collaborations.[14] Among Indian publications, 17.4 percent are coauthored by foreign people. In China, the total is around 15 percent. Given the much higher total publication count in China, their actual number of foreign collaborations is much higher than ours. For the United

Kingdom, the number is 40.6 percent, Korea 25.6 percent, Australia 40.0 percent, and Israel 41.7 percent.

Chinese students, who go to the United States for their graduate studies and then settle there, very often connect back to their parent Chinese institutions. They pursue collaborations with faculty from their undergraduate days. This has the added benefit of providing a pathway for students in China to copublish. Unfortunately, this seems to happen very rarely in India. Undergraduates from IITs who make it into top Western universities for graduate studies and faculty positions seldom connect back to their Indian schools. Two senior professors of Indian origin at US universities told me that they visited their undergraduate institutes in India to investigate some possible partnerships but it eventually came to nothing. They cited the reason as the lack of culture of achievement in Indian institutions.

What Our Researchers Want

I have discussed many issues with the environment for the faculty members. Obviously, all of these are not equally serious. To get the faculty's point of view, we surveyed them on their opinions of the research environment. We gave them a list of 14 parameters and asked them to rate the quality on a 5-point scale. Table 6.3 gives a summary of their responses.[15]

Table 6.3: Average Rating of Research Ecosystem Quality Parameters by Faculty

Parameter	Average Rating
Time taken/bureaucratic hurdles in getting funds or new equipment/machines	2.26
Availability of sufficient competent technical and administrative staff	2.45
Institutionalized ways to interact with industry	2.68
Availability of sufficient funds for travel	2.70
Critical mass of researchers in your field	2.81
Ease of collaborating with/inviting international partners	3.19

Parameter	Average Rating
Departmental/institutional encouragement/incentives for excellence in research	3.19
Quality of PhD students	3.21
Meritocracy in distribution of funds	3.22
Availability of equipment/infrastructure	3.27
Quality of peers and peer interaction	3.33
Salary amount	3.38
Teaching load (considering quality of teaching assistants)	3.48
Campus life facilities—housing/schools/day care	3.55

Note: Based on: Very Poor: 1, Poor: 2, Average: 3, Good: 4, Excellent: 5.

We find that 5 out of 14 parameters are rated below average (<3). On the other hand, no parameter is rated good or above (>=4). The five parameters below average include: access to new equipment, staff availability, assistance with industry partnership, travel funding, and a lack of critical mass of researchers.

On the other hand, respondents seem more satisfied with their salary, teaching load, and campus life facilities. These are rated above average, but still merely good rather than excellent. High achievers will not be swayed by the merely "good" when there are so many other career paths available. Overall, our institutions have far to go to win the confidence of our researchers.

7

Research, Industry, and Science Entrepreneurship

Invention is not enough. [Nikola] Tesla invented the electric power we use, but he struggled to get it out to people. You have to combine both things: invention and innovation focus, plus the company that can commercialize things and get them to people.

—Larry Page, Co-founder, Google

Research creates economic value (and in many cases, public good) through interaction with businesses. PhD graduates go to work in private industry, bringing their academic research experience and expertise, and applying them to the development of commercial goods. Furthermore, businesses often engage university research faculty in consulting roles and sponsor their research programs. Companies even set up their own labs and research teams while simultaneously collaborating with university faculty in various ways. Oftentimes, students and their faculty decide to become entrepreneurs themselves—cutting out the middleman to capitalize on the fruits of their labor, bringing new products and services to market. Thus, for any society that hopes to reap the returns of its investments in education and research, a healthy partnership between industry and the education sector is essential.

Researchers especially in applied fields such as engineering have further reasons to engage with industry. Industry is closer to the customers—the real-time users of technological products who learn of problems and new use cases first. It is a guide for faculty on what problems to address. Of course, this is not the only guide for research. Industry tends to think in terms of a 1–5 year horizon for product development, whereas an academic researcher thinks on a grander scale of 10–20 years. Academics ponder groundbreaking and disruptive innovations, anticipating needs that have not yet arisen. Yet businesses add to the researcher's blind spots and add realism and exposure to technological progress in the industry. In this way, industry and university complement each other to form a productive and mutually beneficial circle.

In the EECS Department at MIT, John Wyatt was my academic advisor and Harry Lee was my instructor in a graduate-level analog design course. Harry busied himself with designing circuits that were immediately useful to industry. He and his students designed several analog-to-digital converters (ADCs) that were considered break-throughs and were immediately deployed for commercial use. At this time, his group was in the process of developing a new compara-tor-based switched capacitor design to eliminate the op-amp to create a large power savings. Meanwhile, John Wyatt was working on an elec-tronic retinal implant to help the blind to see. Although the prospect of a prosthetic eye was fascinating, it wasn't of immediate industry interest because they could not predict a viable product on the horizon, even 10–15 years down the road. However, John had faith, and so he formed some companies to test his research and bring some mature components to market.[1] Similarly, professors who work in quantum computing or new material research are also ahead of industry. Societies need both types: Harry Lee to work in close cooperation with industry to fill immediate needs, and John Wyatt to forge their own path and create shocking innovations with long-term value, taking industry assistance as needed.

Commercialization of Research: Our Performance

We have seen that India does not perform disruptive research. Does our research at least result in some economic value? Are we producing results and products that have some commercial viability?

Let us consider the private sector. Few of our companies engage in research, utilize research, or are what we would call innovative. India ranks 14th in patent-filing activity. The number one and two spots are held by China and the United States. Indian companies and institutions are not represented among the top 100 patent applicants worldwide.

Consider exports. High technology products comprise only 8 percent of our exports. The United States and China export 19 percent and 26 percent, respectively. In absolute numbers, we are less than 3 percent of China and around 11 percent of the United States.[2] Among G20 countries, private research investment on average is 1.48 times public investment. In the United States, the United Kingdom, South Korea, and China, the ratios are 1.97, 1.4, 2.69, and 3.06, respectively. India's ratio is 0.4–0.5. India ranks 9th in total research expenditure in terms of purchasing power parity (PPP) among the G20, and we are 12th in terms of private research expenditure.[3] Our private sector spends much less on research than the private sectors in other countries—an indication of how much our private sector values research.

Our universities are less engaged with industry compared to their Western counterparts. Private industry contributes 19 percent of MIT's total research funding, approximately $128 million.[4] Stanford receives around $300 million, while UC Berkeley receives approximately $80 million. Investment figures for the top Indian institutions are not published, but according to data from IIT Madras, they receive approximately ₹19.51 crores (2015, $11.4 million in PPP terms).[5] This comprises around 9.7 percent of IIT Madras' total research funding. Table 7.1 lists some other metrics of research commercialization and industrial engagement.

Table 7.1: Patent Applications and Technology Transfer Income by University

	IITB	IITM	IISc	MIT	Stanford	UCB	Tsinghua University
PCT/US Patent Applications	16	13	6	341	655	263	500
Technology Transfer/ Royalty Income (PPP, $ million)	2.4	1.07	0.12	62	108.6	15.1	–
Year	2014– 2015	2015– 2016	2014– 2015	2016	2013– 2014	2014– 2015	2015

Source: Annual reports or institutions' communications offices.

The difference in the number of patents is dramatic. The world-class Western and Chinese universities file more than 250 US patents or PCT applications. Indian universities file fewer than 20. Our institutions also file few Indian patents: IITM 18, IISc 26, and IITB 72. Likewise, the royalties we receive on our patents are much smaller, ranging from $0.12 million to $2.4 million. By contrast, MIT and Stanford earn $62 million and $108 million, respectively, on their patents. True, much of this sum is windfall from a few highly successful patents. Nevertheless, it indicates the impact and benefit of the disruptive research that their faculty and students perform. They create not only scientific advancement but also a substantial revenue stream for reinvestment in research and of course financial reward for the inventors themselves.

The Institute and the Industry

We can cite various reasons for the lack of interaction between industry and research institutions. History accounts for much. Before the liberalization of the 1990s, India's economy grew at a slow pace. Industry was mainly traditional, and only a few companies were actually large. Even fewer were scientifically oriented. Therefore, they had neither the interest nor the resources to put toward research endeavors. Furthermore, the companies themselves had little reason to consider innovations to their own operations, given that they were protected by government regulations and had little competition. As industry had little reason to interact with the universities, so did the universities see little of value to be had in industry collaboration. The research faculty preferred to associate with their counterparts in the West and pursue the research problems that the West could provide them. True, many of these research problems were inspired by Western industry, but it was impractical to consider communicating with Western industry directly. The modern communications revolution had not yet begun, geographical distance was unbridgeable, and they had few avenues of introduction.

Following liberalization, Indian industry changed dramatically. India emerged as the IT-services destination of the world and, in parallel, developed a large local market with a new middle class with growing spending power. Companies grew larger and expanded in number to serve both foreign and domestic customers. These were not enterprises of innovation, but they rather worked to their competitive advantage of providing services more cost effectively. However, as they grew richer, they became more interested in supporting research for product development and for innovative ways to grow. In the present time, many of these companies understand that to be globally competitive and to sustain growth, they must pursue research and innovation. They have accepted the increasingly flat nature of the world, where consumers can buy both goods and services from anywhere.

China came to this realization two decades ago and responded by pumping money into their research enterprises, both public and private. However, the Indian institutions did not keep up. The IITs did train undergraduates for the IT industry, but fell short in the fields of science and scientific research. Furthermore, the academic community's skepticism of the private sector is not easily displaced. They remain apathetic to industry and continue to focus on paper publications and citation gathering.

A debate rages as to how best to measure impact. Some argue for citations, which represent an endorsement by the highly esteemed research community that one's work is new and useful. Others argue that industry adoption is the better measure, because it points to economic and/or social value. Industry's critics contend that such deployment is often little more than smart engineering, not true high-quality scientific investigation and discovery. The debate is quite a useless one. Both kinds of research and researchers add value to the society. One is more evident in the short term while the other presents itself long term. Anyway, as we follow the path of a product from concept to production, we see that the two sectors complement each other and need each other. Therefore, institutions must encourage both fields, promote excellence in each, and maintain a respectful and balanced relationship between them. Application of research keeps an institution grounded while pure scientific research charts the path to the future.

Many Western institutions maintain an outreach office to liaison with industry. They investigate opportunities and facilitate cooperation between faculty and private business. They are particularly adept at locating private funding for faculty research. MIT operates an Industrial Liaison Program (ILP) with a professional staff of 50+ people. Their website reports:

1. Over 200 of the world's leading companies partner with the ILP to advance research agendas at MIT (FY15).
2. ILP member companies account for approximately 40 percent of all corporate gifts and single-sponsored research expenditures at MIT.

I have had great personal experience with MIT ILP. About five years ago, MIT ILP partners indicated they wanted to engage more with MIT alumni start-ups. In 2014, I was invited to join an MIT delegation to China to meet a team of government officials and company representatives. This was my first trip to China, and it planted the seed for our Chinese operation. Now we have a

15-person team in China, and we work with some of their top com-
panies. I also received two speaking invitations through MIT ILP.
One came from KPIT Technologies, an MIT alumnus company in
Pune and ILP partner. My advisor from MIT was also a speaker.
She and I collaborated to organize a data science workshop on the
same trip.

But the university–industry partnership in India is not quite
so healthy and productive. In our survey, faculty rated institutional
mechanisms for interacting with industry as below average, the third
lowest among 14 parameters. PhD students rated industry interaction as the lowest among 11 parameters of quality (2.02 on a 1–4
scale). Our institutions do not have professionally run office to
liaison with the industry and explore funding or collaboration opportunities. They do have an industry interaction cell. However, they are
run by faculty, who take up additional responsibility and mostly
respond to incoming queries regarding industry projects rather than
conduct outreach to industry. Their websites provide scant information, are rarely updated, and are many a times disconnected from
the main university homepage. They sometimes seem like dummy
pages put in as placeholders! On the positive side, there have been
some recent efforts to install industrial parks on university campuses,
such as IIT Madras and IIT Mumbai.

While Indian universities do not invest in industry partnership,
Indian industry seems to care for university partnership. Feeling
rebuffed by our own institutions, some of our more mature
industry players have begun to look westward for collaboration.
Not only are these industrialists welcomed there but they acquire
greater brand equity by the association. Narayan Murthy famously
said in 2014:

> [W]e have Indian Institute of Science (IISc) which is hardly about 12
> kms away.... None of them bothered to come to any of the Indian
> companies. On the other hand, the president of MIT, Cornell,
> Caltech, Carnegie Mellon, Cambridge ... you name it and they all
> came to us saying What problems of yours can we solve?

Industry is returning the compliment. Many Indians have begun donating money to the foreign universities. These names include Ratan Tata, Anand Mahindra, and Murthy. The Indian media criticizes them for choosing to sponsor foreign schools,[6] although many of these same people also donated to Indian universities as well.[7] The criticism is misplaced. People apply their money where it will perform best and bring the best return. They give money in order to support a cause or endeavor close to their heart, to sponsor a particular kind of talent, and to realize their vision of a problem solved. Today, people see this potential in the West. Like it or not, Indian institutions have to compete on a global playing field.

Government also has a role to play in spurring innovation. American universities, nonprofit institutions, and some types of small businesses may own inventions that they developed with federal funding. This issue of intellectual property ownership was settled in the United States by the Bayh–Dole Act of 1980. It has been a major driver of innovation. Prior to this legislation, only 5 percent of patents were commercialized. However, the legislation recognized that greater economic benefits could be realized if the developing entities themselves were allowed to commercialize their research results.

India passed a similar rule much later in 2008, known as the Protection and Utilisation of Public Funded Intellectual Property Bill. This sought to be the Indian version of the Bayh–Dole Act. Experts content that the bill still lacks clarity on its applicability. It seems to conflict with the government's general financial rules. Furthermore, it imposes a framework of sanctions against institutions that seem to defeat the bill's intent. Formal IP policies for some of our top institutions have come up only recently, for instance, IIT Bombay in 2003 and IISc in 2004. Being late is fine—but now we need a clear policy that incentivizes all stakeholders for commercialization of their research.

Science Entrepreneurship

Universities also create economic value through entrepreneurship. Table 7.2 displays the country of origin for the world's "smartest"

Table 7.2: Smartest Companies by Country

Country	2014	2015	2016
United States	36	37	32
China	5	4	5
Germany	2	2	2
South Korea	2	0	1
Canada	2	0	0
UK	2	1	3
Israel	1	1	2
Denmark	0	1	0
France	0	1	0
Italy	0	1	0
Japan	0	1	3
The Netherlands	0	1	0
Nigeria	0	0	1
Switzerland	0	0	1

Source: See https://www.technologyreview.com/lists/companies/2017/ (accessed on September 27, 2017).

50 companies, as identified by *MIT Technology Review* (MIT TR). These are companies that "best combine innovative technology with an effective business model." From 2014–2016, not one company from India made the list.[8] In contrast, China consistently has 4–5 in the list. These include Baidu, China's dominant search engine, Huawei, which codesigned and manufactured the Nexus 6P phone with Google, and Tencent, the owner of the messaging app WeChat. Always, the United States accounts for the largest number of companies—more than 60 percent. In 2015 and 2016, more than 50 percent of these "smart" companies were engaged in two fast-growth areas: biotechnology/life sciences and computing/communication.

Despite our strong economic growth and the recent start-up boom, we do not produce innovative companies. Although our start-up models are new to India, similar businesses have existed in the West for a number of years. Consider e-commerce, cab aggregation, secondhand goods classifieds. Large US-based companies have dominated each of these domains. In fact, I believe that as Western companies begin to target the Indian market, many of our copycat companies will forfeit their position. Their early starter advantage will be no match for the super research, technology, and innovation of the Western companies.

We see this already with Amazon starting to outpace Flipkart and Snapdeal.[9] Amazon is a marvel, and you err if you think it is just an e-commerce company. It is a technology company that is pioneering cloud computing, crowdsourcing technology, AI, and service-oriented architecture of software development, while also finding it in itself to produce slick products like the Kindle. India has nothing even remotely comparable to a Google, Amazon, Tesla, or Apple. In contrast, Didi recently bought out Uber's business in China (stated as a merger). Andrew Ng, a celebrated AI professor from Stanford and the founder of the Google Brain project, is now the chief scientist at Baidu Research.[10]

The entrepreneurs who emerge from our top institutions are mostly undergraduates. The enterprises they establish might perform some interesting business innovations, but they do not engage in world-class scientific or technological innovation. Because there is no critical mass of such companies, advice on how to build and scale them is not easy to find.

The ecosystem for science-related entrepreneurship is not well developed. High technology companies may take several years to bring a product to market. For society, the result is usually worth the wait. However, Indian investors tend to think along different lines. Investors prefer to support low-risk businesses with a defined product line and a large preexisting market. They have less appetite for high-risk businesses that require time and resources to develop a commercially viable product or service. Typically, they want to see a product within six months to a year. However, the norm for high technology businesses is usually two to three years before a market-able product is finalized. One of my friends started a computer-aided design (CAD) company in Canada. Three years passed before he conducted his first customer demonstration. In fields such as biotechnology, the time horizon may be even longer.

Another problem for science entrepreneurship is market access. Considering CAD, most customers are semiconductor companies, which are mainly concentrated outside India. Although they might have representatives in India, purchasing decisions are made abroad. Therefore, European and American start-ups have an advantage given their closer proximity to the business customer base.[11] However, in the B2B field, the Internet makes possible many opportunities for domestic producers of goods sold (better, if also used) on the Internet. Two good examples are Freshdesk, a company selling customer support software, and Zoho, which provides cloud-based solutions for business. Nevertheless, sales to large companies typically still happen through face-to-face encounters. Indian companies would require much larger investment to facilitate this.

In comparison, B2C products from India could be sold to consumers across the world, with little incremental cost. There is no reason why a LinkedIn, a Google, or a WhatsApp could not originate or be based in India and address global markets.[12] None of these companies sells or promotes their services by physically meeting their B2C users. This is an opportunity for Indian science entrepreneurs.

We do have examples of some research-led companies from India. For example, faculty from IISc founded Strand Life Sciences, which delivers personalized medical assistance based on genomic analysis. Using this technology, doctors can recommend the most effective treatment based on the sequence of an individual's particular genome. Hence, two patients with the same form of cancer will receive different medicines and therapy, in accordance with their genetic code. They have published 30+ papers on various aspects of their work. Another example is Achira Labs, founded by Dhananjaya Dendukuri, an MIT PhD and a friend. His company uses advanced microfluidics technology for medical testing. They have published their work in four international publications and have received two US patents. They have four additional patents pending. They established their company in 2009, but did not introduce their first product until 2016.

What Can We Do?

Now is an opportune time to spur interaction between universities and industry. Methods and inventions in several fields are now mature to solve real world problems than just being lab prototypes. These can be exploited by businesses to create economic value. More than any time before, the private sector recognizes the value of research and innovation for sustaining growth. They are keen to engage with universities. Many even employ their own research teams.

India of today also represents a new opportunity for researchers, entrepreneurs, and investors with vision and patience. The economy is much larger than ever before. We have many more large companies than earlier, and we have a large and growing market. Private industry now has the capacity to invest in research. We have size enough to capture the economic advantages and social benefits of our research. Moreover, neither our universities nor our industries are confined to each other. With the ease of modern communications, partnerships can be made globally according to where the most interest and opportunity lies.

Five things must be done to make this happen. They are as follows:

1. We need institutional mechanisms for creating university–industry interaction. Universities should go beyond paper counts and citations, and respect their private sector partnerships and the commercial deployment of their research. Universities themselves should establish professionally run industry outreach offices. Such departments must be prepared to look globally for collaborative opportunities, not just locally.
2. Industry and academia must drop their cynicism and distrust of each other. Both must reach out beyond their traditional domains. Research professors must take an interest in practical problems while industry must think beyond short-term gains and quick fixes. As they are engaged in a mutual pursuit,

they require a common vocabulary to understand the other's capabilities and constraints. Each side needs to walk some distance and delay gratification in order to reach the point of partnership and mutual advantage.

3. Successful research ecosystems require proximity. Collaborative partnerships are most evident when industry and academia share geographic space. Accordingly, we should build our new universities near industrial zones. New industrial areas should be marked out near universities.

4. The local ecosystem needs to be developed. In tangibles, science-related start-ups require greater and easier access to capital and good business mentoring and consulting. An ecosystem needs much more—a culture of inventiveness, respect for ideas, collaboration, business savviness, persistence, easy market access, and so on.

5. There should be greater government incentives for university–industry collaboration. This can be in form of tax breaks and subsidies on private sponsored research at universities.

8

Questioning the Existing:
Research Questions That Matter

In the Hitchhiker's Guide to the Galaxy, *Deep Thought*
took seven and a half million years to get the answer to
the question of life, the universe, and everything. The answer was 42!
The answer was worthless, because we do not know the question!
Questions are important.

Today, our researchers do not take up endeavors of original research. We do not seek big, new, courageous questions. Instead, we look outward. Western scientists lay out the questions that interest them most and then lay the foundations. Later, we identify small areas where we can make small contributions. If we wish to do disruptive research and achieve breakthrough answers, we need the courage to ask difficult and open-ended questions. If we do not ask such questions, if we do not dream, then there is no chance for achieving big things!

Surveying the last 60 years, we can find hardly any instances of big questions, ideas, and new ways of thinking coming from India. But there are plenty from elsewhere: seeking to understand the human genome (The Human Genome Project), understanding the basis of consciousness (work led by Christof Koch), framing the big questions of AI (Dartmouth Conference, 1956), and evaluating the impact of development interventions in a statistically sound way (pioneered by Poverty Action Lab, MIT). These bold, open-ended questions have led to a tremendous advancement in knowledge—not only by a single researcher but also in laying out a field of study for future generations to follow. For instance, since 1956 we have made great progress on questions of AI. Today we are experiencing the first effects of AI's disruptive potential, and there is still a huge way ahead. But what if the question had never been asked?

Our questions are smaller and narrower—say, can we make some small improvement to a given algorithm by some mathematical tinkering or by adding some components? Can we apply a given algorithm to a new set of data? But why can we not ask a different set of questions—say, can we design a totally new algorithm that beats the current approach and inspire solutions to new questions beyond the first? Or can we not define a still larger question that subsumes the current question? In 1916, when Mahalanobis[1] presented a math puzzle from the *Strand Magazine* to Ramanujan, he gave a solution to not only the problem in question but offered a continued-fraction solution to the larger class of problems! We need to look at "owning" questions, not merely assisting with answers.

There are more examples from the past. Jagadish Chandra Bose sat in a small lab in Kolkata in British-ruled India and framed his own questions on how plants (as opposed to matter) respond to electrical stimuli, leading to questions of what constitutes life. He set up his own workbench and invented his own instruments when existing instruments would not suffice—such as a mechanical x–y plotter for studying circuit responses. He produced results that challenged and surprised his contemporaries, while adhering strictly to empirical evidence and rational argument. His research continues to surprise and reward us today, as we dig deeper into his work and realize how far ahead of his time he was.

The reason that Bose's work was so consequential and far-reaching was because he asked hard and broad questions. One needs to have "big-picture" thinking to conceive of such questions rather than be caught up with details (which, admittedly, are also important). One needs creativity to think outside the box. New questions require courage and faith in the power of science and human persistence to find answers.

Asking research questions is no trivial task. The words "what is consciousness?" are easy to speak. However, one must translate these words into a question in the scientific vocabulary, where there are hypotheses that may be proved or disproved, either by present methods and resources today or by those that may exist in the future. Substantial work in the nature of consciousness was done by Christof Koch at Caltech, both in defining the questions and in working toward the answers. Consciousness is an example of a "discovery" question. An "invention" question would require that the solution could be objectively tested. As broad as the question may be, yet amenable to evidence based studies, larger still would be the contribution. Sometimes, simply stating a problem as a scientific question solves a big part of the problem. Translating an unbounded problem into a bounded one is hard and needs a mind-set and training of its own. It leads to the discovery of new fields, owning them, and inspiring others to take them up.

Our unique context itself can provide new areas of scientific exploration. We encompass many different cultures, languages, religions, and a civilizational history of more than 2,000 years. We have a huge population, the largest number of poor and illiterate people, lack of infrastructure (including digital), various levels of pollution, and a constraint on funds. Differences in biology and environment present health issues not common in the West. We have our own unique kinds of transportation methods (like road infrastructure), habits of travel (driving!), and other social habits. Would these differences not carry in their heart a different set of research questions? Are we taking advantage of our unique situation to inspire a different set of profound and difficult questions?

Consider, for example, the Ramayana, which has 300+ versions. How fascinating to produce statistical and computational linguistic methods to analyze these documents, attribute authorship, and produce a timeline for composition of parts of the same epic. Here, social science research meets technology. These different versions span multiple languages, each of which have evolved over time. Who but an Indian researcher, steeped in the culture and context, could pursue these questions with greater alacrity? It can tell us a lot about how language and thought evolved over centuries.

Another example is hybrid languages—such as English and Hindi, Hindi and Kannada, and so on—over multiple accents. India is unique in this way. How fascinating to explore the natural language methods and speech recognition methods that such hybrids require? Speaking of hybrids, how would a driverless car work on a road with pedestrians and cars that do not follow traffic rules, and when there is a tonga, a rickshaw, and a cow on the road? How can an emergency response system operate efficiently among India's persistent traffic jams? (In fact, a friend and professor at UIUC is investigating this question right now).

These are some examples of broad and narrow research questions that our national context can provide. These are new questions that can create economic and social value for us, while also advancing global research. If we do not take up some of these questions, they may never get answered—to our loss. But we are uniquely advantaged to exploit them as a strength. Our unique environment and ecosystem not only provides us with unique questions: they make possible certain experiments that would not be easy elsewhere. One example is Pawan Sinha's Project Prakash, where he treated blindness of children to answer questions of human vision (see "Project Prakash"). The experiment couldn't happen in the developing world because they don't have treatable blind people. There are other examples, which don't necessarily dwell on our deprivation.

We do not find a lot of Indian researchers asking these questions. We get our questions not from the environment around us but from published research papers: derivative problem statements coming from the West. One could ask why the work being done at the MIT Poverty Action Lab did not first happen in India instead—the country with the highest number of poor.[2] Indian researchers did

Project Prakash

Humans can effortlessly distinguish between a tree, a car, a bus, or a person, on the road. For machines, this simple task is extremely difficult. How does the brain do it? To find out, we could run experiments on babies, at the time when they are first learning to see. However, ethics do not allow us to run elaborate experiments involving babies. Anyway, babies cannot express what they are seeing.

Professor Pawan Sinha, a professor of Brain and Cognitive Science, found a unique way to address this question. He calls it Project Prakash. I first met Professor Sinha in 2005, when he was just launching this project. The idea was simple. He would restore the vision of people with treatable blindness and then run various experiments as they gradually recover their sight. Experiments include scans before and after restoring vision, the reaction of the subjects to various visual stimuli and other means. He conducted this experiment in India as part of a humanitarian project to treat the blind.

His experiment has led to many remarkable discoveries. First, his findings contradict the belief that the brain, with age, loses the plasticity necessary to learn to see. Sinha found that people blind from birth can learn to see in their teenage years and even later, although their restored vision might not be perfect. His discoveries not only have great scientific value but also an immediate health value to the society.

Sinha also found that motion plays a significant role in helping the human brain recognize objects. A person may not be able to easily perceive a triangle if it is overlaid upon a circle. But if you set them in motion—the triangle and the circle in different directions—the brain learns very fast! There are many implications for this discovery also—especially for computer scientists who mostly train their object-recognition software with static images.

Sinha's experiment is a great example of finding new approaches to answer long-standing questions in vision and neuroscience. His "experimental bench" is a gold mine, where new experiments result in several new answers. Sinha's work has earned more than 8,000 citations.

not begin to take notice of technological and econometric methods for development until Western researchers initiated them and gave the field respect. Westerns had to find interest in our problems to get us interested! One area where we may be making some original progress is frugal innovation. Yet perhaps for good reason, Ratan Tata gave a huge grant to MIT (instead of an Indian school) to start the Tata Center to develop the next generation of technology solutions for the developing world. There are other significant research questions deserving our attention that our community has yet to take up. For instance, what is India's historical contribution to world science? How do we improve scientific research in India?

In this chapter, we will investigate why we fail to ask broad original questions and develop new ways of thinking. Part of the problem lies in the lack of proximity of our researchers to the problems in the real world. Do they have the courage and motivation to ask new research questions inspired by these problems? Does our education system promote developing intuitive thinking and hand-on approach that is needed to investigate such questions? Finally, does our research culture encourage and respect such original endeavors? Let us answer some of these questions.

Motivation—Raise the Bar

One of the reasons for our lack of originality is our understanding of academic success. Achievement for us is measured by the number of papers published in top journals and conferences. A PhD student at a top Indian institution is expected to publish 3–4 papers in top journals or conferences to receive his or her doctorate. It is said that for faculty members, the number of papers published and the number of doctoral students guided are the key criteria examined for determining promotions.

Counting papers is not a bad criterion. However, one needs to see beyond it. MIT's criteria for faculty hiring is not so much the number of papers, but includes the vision of the researcher and the power of his or her ideas. To receive tenure at MIT, you must be "either the top investigator in your field, or one of the very tiny handful of top investigators in your field, in the world." Investigation implies much more than paper output. We need to raise the bar. Our aim is not only to get top papers but new and bold ideas.

It is like Vivekananda said, on the need of a religion, "It is fine to be born in a temple, but not fine to die in one." The same is true with regard to publishing top papers. For legitimacy, papers are a good indication that your research training and methods live up scientific scrutiny. But papers should be a starting point, or something ancillary to investigation. Making them a goal is a disaster.

Also, we must remember that the number of papers is not always a good measure of a researcher's contribution. After all, one breakthrough paper outweighs many incremental papers. Shannon's paper on information theory,[3] which forms the basis of most modern digital communication, is one such paper. One must also remember that bolder questions are riskier: whereas one might solve a bounded problem and get a quick paper, one might labor on a new open-ended question for some time, without seeing great progress. Thus, rather than just counting papers, we need to see the work of our researchers from a grander and more nuanced perspective.

If paper is not the goal, what can be alternate motivation? The best research comes when the researcher approaches an unsolved or previously unsolvable question, out of deep personal interest. His or her motivation might arise from intense and enigmatic curiosity (or inventiveness) in a given subject, say, prime numbers. It could be Raman's interest in why the sea looks blue, or Newton's interest in how a general principle can explain both why an apple falls on earth, or moon rotates in its orbit. It could be Darwin's interest in how the varied life forms came about; Ramanujan's interest in what value a continued fraction approaching infinity takes; or Bose's interest in the plant nervous system. To many of us, these may mean nothing, may seem not important, or simply too hard to think about. For some of the world's greatest scientists, such questions are the mission of their lives.

Curiosity is one driver of research questions. Public interest and industrial interest are others. These would include Bose's interest in how to communicate messages wirelessly; Karmarkar's interest in how linear programs can be solved efficiently; Esther Duflo's interest in how to reduce poverty; or John Wyatt's in how to help the blind to see.

We need to nurture scientific curiosity in our next breed of researchers—inspire them to pose courageous new questions, to work on developing new methods, and to seek ways to disrupt current ways of thinking. We limit their horizons by giving closed-ended tasks for the purpose of quick publication. They need to "live" their problems to make breakthroughs. This needs a cultural change.

Scientific Identity and Pride

Jab zero diya mere Bharat ne, duniya ko ginti tab aayi, translated as "The world learned to count, when India invented zero," a popular song from the movie *Purab aur Paschim* (1970).

Do we, as a nation or a race, have pride in our scientific tradition? Did we have a great tradition? Do we know about it? Did our ancestors ask original questions and develop their answers? Do we have role models? Let me tell you a story to get to some of these questions.

In fourth grade, I read in my Hindi book that an Indian scientist named Jagadish Chandra Bose first found that plants have life. The chapter detailed how the British gave lower salaries to Indian professors compared to British professors in Presidency College because the Indian professors did not involve themselves in research. Bose rejected the lower salary. Impressed by his breathtaking research results, the British gave him a salary equivalent to the British professors after two years.

The book also described how Bose had demonstrated a radio wave that traveled over a mile to ring a bell at a distance. But the following year, I read in our science book that Marconi had invented radio. Here was a contradiction that did not escape my young mind.

This question stayed with me. In the final year of my engineering degree, we were required to give a seminar. I posed the question: What was Bose's contribution to the invention of radio? In my review of the literature, I found that a researcher named Probir Bondyopadhyay had discovered that Marconi had actually used Bose's "self-recovering coherer" to receive radio waves. I read Bose's original papers myself and I admit I became gaga over these early devices—coherers and autocoherers. In my seminar and in the article that I subsequently wrote, I demonstrated electrical models of these devices using contemporary circuits.[4]

My own view, shared by many, is that radio was invented by Bose and Marconi both.[5] Do not get me wrong: I do think Marconi made a significant contribution. The crazy big idea of sending radio waves over the Atlantic—and then making it happen—is no mean feat. However, he may not have been successful without Bose's detector. Thus, both deserve the credit. Unfortunately, even today our books report that Marconi alone invented radio!

Life came full circle in 2010, when my friend Gaurav Gandhi and I discussed memristors and switching resistances. He showed me a graph representing the electrical properties of switching resistances. I felt I had seen similar graphs somewhere. Where else, but the work of Bose on coherers and autocoherers! We found these devices had memristive properties. Bose was the first to observe, alone among his contemporaries, bipolar switching of the device. My earlier electrical model of these devices was in fact wrong—they were not a diode with a capacitor, but a memristor! Bose even had an explanation for the physical process behind the switching behavior, which matched some we have today. Another scientist of the time, Ecless, had a mathematical model for the physics of coherers, which had striking similarity to Chua's equation on memristive systems that came out in 1976. By digging in Bose's work, we provided the world with the first discrete memristor, and also a characterization of its switching behavior. This possibly constituted the canonic implementation of a memristor.

This long story is to illustrate several issues of interest to us and draw some very relevant learnings. First, India's historic contribution to world science is not sufficiently studied, documented, or popularized. In truth, this is a common topic of conversation, but unfortunately it is discussed with little true insight. The right wing claims that ancient India was familiar with all modern science and technology: we could transplant heads, we had nuclear weapons, we understood the structure of atoms, and we even had airplanes. We cannot disprove these claims without a time machine, but they do not hold up well against what we would consider scientific proof. The left wing on the other hand thinks that the history of India is merely that of Brahminical tyranny, class struggle, and superstitions. As a result, much of it is not there in the academic literature or in our school textbooks.

The undeniable truth is that India has made significant and illustrious contributions to the field of science, which have been consequential for world culture. For example, there is considerable evidence that modern numerals as we know them and the concept of zero came from India. We know that the first written statement of the popular Pythagoras theorem is in Shulba Sutra of Baudhayana around 800 BC in India. The first recorded rigorous proof of it came from China. The Chinese call it the Gougu theorem, based on the name of its inventor.[6] Panini, who systemized the rules of grammar for Sanskrit, was the world's first linguist. All of these asked original questions and provided breakthrough answers and a foundation for future science. We need a proper dispassionate study of original sources to learn about these, which will require a collaboration between scientists and historians.

Why should we do this? First of all, to set the record straight. We must apply scientific rigor to the history of science just as we apply scientific rigor to science itself. Researchers of the past should receive due credit for their discoveries and inventions. History demands accuracy. The world needs to know how modern science came about, how different cultures considered it, and who discovered what. Moreover, India has the responsibility to tell the Indian story, albeit dispassionately and scientifically so that it serves as inspiration to a new generation of great thinkers.

A second important reason is scientific pride, national pride, and the need for role models. We can help the younger generations acquire a scientific temperament by acquainting them with what science has achieved in the past. Our students should be taught about our scientific luminaries at an early age. This reaffirms their belief in the power of science. Realizing how seemingly impossible things became possible instils an appreciation for the labor of the scientist. If these impossible things have been accomplished by a person that a student can identify with—by race, nationality, or gender, for example—the student begins to believe that he or she himself or herself could be the initiator of fantastic discoveries. It is similar to how one Abhinav Bindra would spark a thousand more shooting experts.

Last but not the least, a very important reason for studying India's historical contribution to science is the possibility of rediscovering lost knowledge that has the potential to inspire new knowledge. Scientists such as Tesla, Schrodinger, and Einstein have drawn inspiration for their work from philosophical (and not strictly scientific) texts of India. My friend, Shailendra Mehta,[7] tells me that 30 million manuscripts are known to exist in all the Indian languages of which the National Mission for Manuscripts has catalogued about 4 million. Who knows what further wisdom is contained in these texts?

To sum up, we need to build a scientific identity for Indians. Pre-independence, the Mahatma established an Indian identity based on ethics and morals at the expense of technology (not necessarily, science). Around the same time, Vivekananda created a spiritual identity for India and said that it needs to be combined with Western science for the progress of the world. Today, one dominant narrative is of the computer-savvy Indian. A careful study of our scientific past will help us to create a new identity—of the scientific innovative Indian—coexisting with the others. This will inspire all of us, give us the courage, to go on our own independent search for truth, and be successful.

Complete Solutions to Relevant Problems

We need more collaborations with industry. We should focus more on projects rather than individual papers to think of interesting problems and get a useful output.

—A PhD student in our survey

Our researchers do not pick relevant problems from our local environment to create new research questions. Our local environment presents no end of problems into which our researchers could conduct inquiry and investigation. We find such questions in the things that we contend with every day: traffic jams, pollution, the illness of family members, relatives, and friends, or observing the troubles encountered by household help when they attempt to use a mobile phone. Research-worthy questions can come also by interacting with industries that have close proximity to customer problems. As India houses several global companies today, they can provide not only local problems but also access to questions with global appeal. This can provide a first and direct access to many relevant and pressing global problems, rather than constantly relying on derivative versions in Western research papers.

Proximity to one's problem provides an edge in understanding the problem and discovering new research questions not apparent to other researchers. Not only does it provides solution to problems relevant to us, it is an opportunity to provide new interesting research questions to the world.

One reason why we do not presently interact with industry for research study is the lack of institutional mechanisms for interacting with industry.[8] Still another reason is in history. Until the economic liberalization, our researchers did not have sufficient opportunities or resources to deploy solutions for research problems. Our industry was underdeveloped and short of capital; government support was limited by means; and the innovation ecosystem was essentially nonexistent. At that time, the best we could do was passively receive research questions posed by our Western brethren and then contribute some support to minor areas of their deployment. We could neither identify the questions ourselves nor could we deploy the results of the work. Today, even though we have become the world's third largest economy and enjoy a large start-up ecosystem, these habits of past times remain.

Another issue is created by the fact that our researchers do not work on real problems. By not conceiving of and owning the original problem, we do not take a system's approach toward the solution: developing all the components required for deployment and solving the research questions that the problem poses. Examples of a total system include the building of a driverless car, a cheap laptop, or a chip for testing blood sugar. Or it could be the development of a new kind of prosthetics, which not only help the especially abled but also help answer the questions of human vision and neuroscience. In contrast, our researchers work on narrower theoretical problems, say algorithms to sort numbers, oscillator design, or solving the math of a control system design.

A system's approach brings the research closer to real deployment and tests how different components contribute to solving a larger question. The larger question is less bounded, less clearly expressed in explicit scientific definition or math, and hence comprises a storehouse of more challenging issues. The interaction of different components brings about fresh complexity and requires more rigor in designing the individual components. Today, these interactions could be of very different components—electrical, mechanical, computational, biological, and even social—leading to multidisciplinary research. Multidisciplinary research is in vogue. A lot of contemporary research questions entail multidisciplinary answers beyond those that can be provided by narrow questions in individual fields.

Building a system also helps in identifying new research questions. Such activity provides a substrate or a bench (in computer science terms) that continuously provides new research questions and tests solutions to those questions. The researcher may initially understand questions regarding some components well, while others are less familiar. As she keeps an eye on building the complete system, she works on defining more parts and identifying the problems inherent in them. Furthermore, new research questions may just come up through the interaction of different parts, which were hitherto unthought of.

We also find that building systems is not always respected in the Indian research ecosystem. Many people believe that building systems is not research. Many believe that true research is working on bounded problems, ideally mathematical, with a clear metric of effectiveness. This debate is not unique to India, though it may be more exaggerated here in its partisanship. I heard the debate in the United States also, particularly when groups at different universities were building the driverless car. In my view, both kinds of research are much required and have demonstrated their merit in advancing the goals of research. More recently, given the multidisciplinarity of the problems we seek to solve, systems have risen in importance. In India today, we focus very little on systems. We need exponentially more of this work if we are to take our research ahead.

My research group at Aspiring Minds solves real-world problems encountered in the labor market. Let us consider just one of these, that of accurately measuring skills, to illustrate building systems for real problems. We thought deeply about how this could be done using machine learning and the associated research questions. We came up with a lot of issues and scientific questions because we were thinking of real problems we wanted to grade. We presented these in a paper at the Workshop on Data Driven Education at NIPS 2013. Furthermore, we co-organized a couple of workshops bringing researchers together to think about the long-term visions for assessments, goals, and to define what comprises state-of-the-art solutions. These activities resulted in an interesting white paper,[9] a resource for everyone in the field.

One problem that we worked on quite a bit was how to automatically grade programming skills. This is a big need to help students get feedback on their programs and for companies to hire good programmers. We developed a method to automatically grade programs that mimicked the process that human evaluators would follow to grade these programs. This was a new question for the research community. Our approach was a unique way of combining program analysis methods and machine learning—a field which is new and hot. A paper based on this work was published in KDD in 2014.

Given that the solution was meant to be deployed in the real world, we soon ran into another problem. Our method required that we build a model for each programming question—meaning that we had to collect and manually grade data for each programming question. A real deployment needed to handle hundreds of questions. Repeating the process for each question was time consuming and expensive. To simplify this, we created question-independent methods. This helped our system "super scale" and led to another KDD publication. This method can be used for other grading problems as well. Another associated problem that we are working on right now is how to grade uncompilable codes. And so it goes on.

This example highlights the many benefits of solving a real-world problem using a system's approach. Systems find research questions, define the framework for thinking about the questions, find innovative solutions to particular problems, inspire more new questions, and continuously work to solve them. We have similarly worked on questions of the automatic grading of spoken English.

Our researchers' lack of proximity to real problems, formulation of original questions deter them from engaging in building systems. For some, this doesn't constitute research. The systems approach takes more time, effort, and discipline than the approach that leads to quick publication. But in the long term, attention to problems and engaging in the systems approach lead to "ownership" of the questions, the development of actual solutions, greater fame and recognition, and greater opportunities for follow-on research. None of this gets recognized if one is simply counting papers and citations. As I said before, we need to raise the bar.

Education Culture

I have talked about asking broad open-ended questions versus working on narrower bounded questions. I will expand on this theme with two related ideas: (a) hands-on approach versus theoretical analysis and book knowledge, and (b) intuition versus math. Somehow, India today is biased toward the latter in each category. There could be many reasons for this that are not easily substanti- ated; they may be rooted in our education system. Let us try to understand these a little better.

We can observe an interesting divide in India. Researchers in our labs work on theoretical questions. They are somewhat removed from the society at large and are therefore not well motivated to address societal issues. Concurrently, there are people in the society who do work on these questions, driven by necessity or by business opportunity. Mostly, these latter folks do not have research training. Thus, they conceive of hacks (famously called *jugaad*) to solve the problems and serve the needs they identify. Usually these *jugaad* consist of some smart engineering tricks, which succeed in solving the problem to some degree, in a limited way. While some of these *jugaad* are quite clever, they mostly do not advance our understanding of the research question or lead to fundamental improvements in developing new solutions. This leaves us in an interesting situation: those who have research training do not access (or have access) to the problems and those who address problems do not have research training. If these two could combine—our researchers have proxim- ity to problems and problem solvers have research training, it would pay rich dividends.

Unfortunately, we are accustomed to the hands-off approach. Let us consider school-level math education. We teach theoretical con- cepts like solving linear equations (two equations with two variables) using a step-by-step math procedures. Teachers rarely use activities and real-world examples to motivate the question. One example of an engaging hands-on activity is distributing blocks of different weights and lengths to the class and then instructing the students to arrange them in a line in a box to achieve a given weight. Then let them find the right combination by hit and trial. Undoubtedly some of them will get unnerved by the difficulty of the task and appreciate the beauty of math in solving the problem. Others might find hacks and limited solutions to solve the problem—it would spark their imaginations and engage their creative selves. Yet still others might reveal themselves as a Ramanujan or a Bhaskaracharya, astounding the instructor by identifying the general problem underlying the specific project and finding the solution to it!

The same project could then be used to discuss how people have solved this problem historically. That may lead to more ideas than our current solution approach involving variable elimination. The stories of the scientists could be a further motivation. An Indian name might just pop up as well! With the same example, by a few modifications, one could motivate advanced topics such as linear inequalities and linear optimization.

In contrast, we teach dry equations and step-by-step, problem-solving procedures. We test students on whether and how accurately they can follow procedure. Math instruction does utilize word problems, but they are far away from reality: mostly they deal with ages, reversing numbers, and other so-called real-world problems that never actually occur in reality. Such lessons do not foster a problem-solving approach in our kids, spark their creativity, or convey the usefulness of mathematical techniques in the real world.

The example I just gave is of an artificial problem created to motivate a mathematical principle. We can better experience the real world by observing our surroundings, defining real problems, experimenting with solutions, and building projects. For example, alongside textbook readings of the various kinds of soil used by farmers, students could conduct analyses of the soil in the school field. Similarly, one can do more than read about flora and fauna in African jungles: observe the many specimens in nearby parks. Walking amidst the pollution in the city, sitting in traffic to better analyze it, moving about the slums to better understand the challenges of poverty—all of these activities will spark questions and impart valuable lessons and concepts. On the invention side, students could seek to build their own 3-D printer, a toy driverless car, a drone, a blood pressure machine, or even an app to detect cancer through images. Any activity that engages the hands as well as the mind[10] will teach one to apply multiple concepts for constructive ends.

We should learn to combine our traditional focus on theoretical concepts, processes, proofs, and formalism, with a more hands-on approach to problem solving. As a graduate student at MIT, I generally found the Indian students to be better at the theoretical work, while the students from other countries were hands-on at building projects. In my understanding, India's historical past has been more about "discoveries" than "inventions." We need to balance these.

Next, let us consider the benefits of intuitive understanding of a problems as opposed to merely being versed in the mathematical equations underlying it.[11] In the above example of linear equations, there is a math process to solve it by eliminating the variable. What is its intuitive understanding? It might be a little complex in this case for a school kid, but here it is: We are looking for numbers that satisfy both equations. There are many pairs of numbers that satisfy one of the equations. We can simply find them by assuming any value for one variable and then calculating the other using the equation. Thus, the first equation makes one number a "function" of the other: for every unique value of the first number, we will have a unique value of the second number. If we can replace this function in its symbolic form in the second equation, we will be done. Such an intuitive understanding, a higher level thinking, gives new insight into the problem and allows one to apply the reasoning to other problems as well.

Similarly, consider amplifiers, a subject taught in twelfth-grade science class, and then in the first two years of electrical/electronic engineering. Amplifiers are present in almost every chip in the world. Why does the gain of the amplifier go high, low, or medium, when you use the transistor it in different orientations (common emitter, common base, or common collector) or when we attach a resistor to the emitter in the common emitter position? This can be worked out by systematically writing all circuit equations and calculating the gain. On the other hand, it could be understood intuitively, by imagining the behavior of the transistor, resistors, and how they affect each other. With such an understanding, one can answer the gain question without solving any of these equations. Similarly, the reason why a feedback loop stabilizes gain and reduces noise could be expressed through equations or understood intuitively. Most top analog designers rely on their intuition to understand circuits and design new ones. Rahul Sarpeshkar, a top analog designer at MIT, greatly stresses intuition in his low-power analog design class.

Math equations and intuition are both important to a researcher. Some can think through equations, while others achieve breakthroughs through intuition. Those who make the most progress are the ones who can deal with math equations and also intuit what is going on. In India, we find more emphasis on the mathematical procedures and understanding. We put less emphasis on intuition. Intuition helps one to think abstractly about things, and to see the big picture and serves us better in open-ended problems, by paving the way until something is ready to be mathematically expressed. We need a larger emphasis on intuition in our teaching and research training methods.

These three notions—broad open-ended problems, the hands-on approach, and intuitive understanding—connect with each other. The government's recent initiative to establish the Atal Tinkering Labs in schools and institutions is a great example. These labs, some of them already established, contain DIY kits, 3-D printers, microcontroller boards, sensors, and computers. Students use all these resources to become builders and innovators, and to solve open-ended problems. Mentoring would further help develop identify pressing problems, big-picture thinking, and intuition. Professor Tarun Khanna[12] recently described to me how a school student in one of these labs developed a robot for drip irrigation!

In 2015, my colleagues and I took it upon ourselves to figure out how to teach data science to fifth-grade to eighth-grade students. We think that data science is becoming a horizontal skill, very much like computer literacy, and all jobs in the future will require some aspect of it. We used all three principles in designing the course material. Rather than deliver passive lectures, we directed students to actually solve a problem and make a predictor themselves. We chose the problem of making a "friend predictor," which the kids found fascinating. We designed a hands-on experiment wherein students were given flash cards with details of people and they had to decide for each, if they will make them a friend. Following this, they worked in groups, entered data in excel, made graphs, calculated numbers through excel formulas, and wrote about their experiences in a blog. One goal throughout was the development of intuition in the kids. Therefore, complex data science models were simplified to intuitive ideas; graphs were interpreted in groups to derive insight; and students were motivated to think of new "features" to determine whom they would befriend and whom they might not. We presented our pedagogy to teach data science to kids, a first such attempt, in SIGCSE 2017.[13] This is one example of teaching using hands-on exercise and developing intuition.

We need to revamp our education system and research training ecosystem to imbibe some of these ideas. To complement our strength in theoretical analysis, we must become more hands-on and develop intuition as a thinking and research tool.

9

Building World-class Universities: Policy and Institutional Structures

Research Universities are considered among the central institutions of the 21st century knowledge economies.... The few scholars who have attempted to define what separates elite research institutions from the rest have identified a number of basic features—high qualified faculty; excellence in research results; quality of teaching and learning; high levels of governments and non-government sources of funding; international and highly talented students; academic freedom; well-defined autonomous governance structures; and well equipped facilities for teaching, research, administration and often student life.

—Philip Altbach and Jamil Salmi,
*The Road to Academic Excellence: The Making of
World-Class Research Universities*

Great universities do not exist in a policy vacuum. The majority of the top research schools in the world are public institutions. Even private universities get the majority of their research funding from government sources. Hence, much of the progress of research is influenced by government policy in science and technology, higher education, and areas such as defense.

Policy can be a great enabler. It can make resources available, create a market, and spur institutions to compete and grow. The Indian economic liberalization of the 1990s is a great example, taking India's economic growth from 3–4 percent to 7–8 percent.

What policies could further the development of Indian science and technology? How has the government thought about this over the last 60 years?

Second, we wish to understand what structures are required for a twenty-first century university to function efficiently? We have already discussed the centrality of people, the right environment, and availability of resources. These are important, but are there other components needed?

The Web of Institutions

University and Government Labs

The majority of research is performed in university and government research labs. In the middle part of the last century, most countries had a larger concentration of researchers in government labs. However, in the last few decades a larger portion of research has shifted to universities. Table 9.1 shows the ratio of government research personnel to university personnel over the past 40 years.

Most countries have witnessed a shift in research personnel from government to universities. Other than China, where the ratio of government personnel to university personnel approaches 1, all the rest are above 4. In the United Kingdom, there are 20 times more research personnel in universities than in government labs.

India is a clear outlier. Our university research personnel is one-fourth our government research scientists. India like many others made a decision in the 1940–1950s to park our research in government labs. Universities were expected to teach learned knowledge, not create new knowledge. Scientific research was left to large labs such as the CSIRs for industrial research, and DRDO, ISRO, and Bhabha Atomic Research Centre (BARC) for atomic research, among others. Whereas some of these institutions have had remarkable achievements such as India's Mangalyaan Mars orbiter and CSIR's work in chemistry, other labs have not been so productive.

Table 9.1: Government and University Research Personnel by Country over Years

		USA (1981–2002)	China (1991–2014)	UK (1981–2014)	Australia (1981–2012)	India (–/2010)
1980s–1990s	Government personnel	59,200	200,700	20,000	6,794	–
	University personnel	98,300	132,700	25,000	13,610	–
	Ratio (University/Government)	**1.66**	**0.66**	**1.25**	**2.00**	**–**
2000s–2010s	Government personnel	47,371	295,899	7,640	8,311	87,905
	University personnel	186,049	282,304	158,491	65,772	22,100
	Ratio (University/Government)	**4.00**	**0.95**	**20.75**	**7.91**	**0.25**

Source: Data from OECD, as per availability. Data for India from Department of Science and Technology.

However wise such a decision might have been 60 years ago, the continuance of this research policy is a tactical mistake. The time has come to redirect our resources away from the traditional large government labs and put it into university-based research. The model of driving research through doctoral students, combined with teaching, has paid great dividends globally. It creates a pipeline of newly trained researchers. It will earn us a greater return on public investment through greater research productivity and also through the spillover effects on teaching and entrepreneurship. But are the universities ready to take on these responsibilities?

The University System

India has more than 300 university-status institutions. These include (a) state, central, and deemed-to-be universities, (b) institutions of national importance, and (c) other government and private colleges. India borrowed from the British university model, wherein most state and central universities oversee a network of affiliated public and private colleges to provide guidance, conduct examinations, and to ensure that the schools operate according to government regulations and academic standards. They have small teaching departments and student population. Their emphasis is not research. Today, India is one of the four countries that still have a university–college affiliation system. Each of these are or were commonwealth countries.

Next are the stand-alone institutions of national importance, such as the IISc, IITs, TIFR, and AIIMS. They provide first-rate undergraduate programs, the graduates of which are generally in high demand in the employment market. They attract people who want to become professionals such as doctors, software engineers, and consultants. They do not necessarily attract people with scientific ambition. Yet some of these are our best research institutions. Unfortunately, no private institution does well in research, even though some of them do well in teaching.

These institutions are few in number and they have neither the size nor scale to do world-class research (refer Table 9.2 later).

They tend to specialize in just two–three disciplines, which denies them the opportunity to do multidisciplinary research and cross-fertilize ideas. Some of them concentrate on teaching and do not prioritize research. The institution leaders occupy themselves with administrative matters, rather than pursuing a vision to build the school's research potential. The IITs have begun investing in research over the last couple of decades, but often find themselves in a dilemma about their core mission and responsibility for the future.

Beyond the universities and stand-alone institutions, India has some 15,000 colleges. These emphasize teaching, not research. However, they have an important impact on the research ecosystem. These colleges are feeders for good PhD students. Second, they are end destinations for many PhD students; following their doctoral studies, many return to the colleges to teach. So on the one hand, they need to instill curiosity and passion in their students so that they might go on to doctoral work, and then they must be able to provide professional satisfaction in order to make themselves an attractive career option for the PhD graduates. These two are symbiotic.

Unfortunately, the bulk of these colleges do poorly in this PhD preparation. First, they do not provide very good training. Aspiring Minds' National Employability Reports have determined that fewer than 10 percent of engineering graduates possess domain skills sufficient for a typical job in their fields. We find that only 19.6 percent of B.Com graduates have basic accounting skills. For some other fields, the statistics are even worse![1]

A key reason for these poor student outcomes is poor instruction. A large proportion of faculty at these colleges are those who were not able to find a job in the private sector. Academic professions do not pay very well and have low social status. Many institutions pay less than the salary mandated by the pay commission using corrupt means. The perspective of the university administration and the interest of the students in quality training are suspect. This is a huge issue in itself, but beyond the scope of this book.

For our purposes, suffice to say that these institutions are not an attractive destination for highly skilled and well-trained PhD students.

Our universities are more affiliating bodies than research institutions. There are individual institutions of excellence which do not do world-class research, and our teaching institution ecosystem doesn't work well.

If the universities are to be the future flag bearers of research, they need reform.

Missed Opportunities—The History

Jawaharlal Nehru, our first prime minister, had a great vision for India's scientific and technological future. He thought that science would play an intrinsic role in the development of the nation, and so he devoted himself to education policy including the establishment of the IITs. His vision was not misguided. These IITs and other like institutions have paid enormous dividends to our country. We see the benefits in the IT industry alone.

Given the early emphasis on science and education, why has research performance been disappointing? Part of Nehru's policy was to put science and technology in one box and education in another. Nehru's government created a large web of government-run scientific laboratories around India that were meant to do research. The IITs and other institutes were conceived to train professional manpower for the industry. And the universities were appointed as overseers and examiners of the college system, much like the University of London. The Education Commission of 1948–1949 identified teaching as the primary function of higher education, with barely a mention of faculty's obligation to research. Similarly, the science policy resolution of 1958 made no reference to higher education.[2]

The Kothari Commission of 1964–1966 presented a big opportunity for the future of scientific research in education. Among other things, the commission recommended the establishment of five or six major universities, "where conditions may be provided, both as to staff and students as well to necessary equipment and atmosphere, to make first class postgraduate work and research possible." The proposal found strong support from F. Seitz of the US Academy of Science and P. M. S. Blackett, the president of the British Royal Society. The commission recommended upgrading some of the current institutions to international universities. They also prescribed the establishment of clusters of "centers of advanced studies" at universities to promote research, collaboration, and multidisciplinarity.

The Vice Chancellor's conference and the Parliamentary Committee rejected the proposal. The prime argument against the proposal was that it was elitist, discriminatory, and against the egalitarian principles of our socialist democracy. They missed or chose to miss the point which the Commission had argued: that a discriminatory approach is necessary for excellence and that the equal provision of resources irrespective of quality and potential leads to mediocrity. The idea of "centers of advanced studies" was accepted but not executed with any fervor. This was a historic mistake! If the proposal for the major universities were accepted and acted upon in the 1970s, India might be on a much different trajectory today.

The Ministry of Education issued a report in 1985, recognizing the issues that had resulted from the separation of research and higher education. It noted that "the major national inputs that have gone to the laboratories outside the universities" have led to a "great deprivation in terms of facilities for frontline work" for the universities. The 2003 policy on science explicitly talks of the role of higher education in the development of science.

In all these intervening years, the government has tried various initiatives aimed at spurring research in academia. These include allocating differential resources among institutions and individuals. Other efforts include interuniversity centers to provide common high-quality facilities, contingency grants for research, expansion of scholarships, funds for international visits, and teaching breaks for faculty to pursue research projects. However, most of these initiatives were "Band-Aid" fixes not backed up by actual changes in policy.

The last two decades has seen a new set of institutions intended to promote scientific academic research—the National Institute of Science (NIS) and the IISERs. In addition, several new IITs have been established. However, many of these institutions are beset by the same issues as the earlier ones.

In 2007, more than 30 years after the Kothari Commission, Prime Minister Manmohan Singh announced his intent to establish 30 world-class universities. For nine years, there was little movement. In 2016, the proposal was revisited, but revised downward to 20 universities. Having missed so many opportunities before, how India will move forward now is anyone's guess.

Inside the Research University

A university anywhere can aim no higher than to be as British as possible for the sake of undergraduates, as German as possible for the sake of the graduates and the research personnel, as American as possible for the public at large—and as confused as possible for the sake of preservation of the whole uneasy balance.
—Clark Kerr, 1963, President, University of California

The research university is a complex organism. Its various parts evolved over time to give it its modern shape. The research element in education first took shape in Germany in the nineteenth century. Germans adopted the Humboldtian model combining teaching and research. The students would learn from the faculty in lectures. They would create new knowledge in laboratories and through seminars. It was a strict hierarchical and authoritative structure to organize the faculty and students as apprentices to their masters, the faculty. However, they did not have large-scale undergraduate programs for technological and industrial training.[3] This is where the British pitched in. They taught the world how to organize undergraduate college education. Oxford and Cambridge were the first large residential undergraduate institutions of the modern world. They had their own endowments, space, faculty, and libraries—providing liberal education through tutorials.

The American university fused the German and British models: large-scale undergraduate programs combined with research by PhD scholars and faculty following their own innovations. However, they discarded the strict hierarchical system of the Germans with a liberal "equal partnership" between research students and faculty and a participative model of shared governance. The Americans also brought to the university an element of competitiveness in pursuit of the largest endowment, the most talented faculty, and the most promising students.

In the latter half of the nineteenth century, society had begun to regard science as a basis for knowledge rather than say, theology. Naturally, this development took place concurrently with society's growing recognition and appreciation for the economic and social value realized through industrial applications of new knowledge created through scientific research. Therefore, university education, which had been dominated by divinity, law, and medicine, now underwent a shift toward science, technology, and engineering.[4]

Unlike the German research university, which disregarded the practical applications of their research, the US university emphasized service to society as a key value. The government provided federal land for the establishment of public universities and experimented with hybrid models of public–private partnership. Industry and industrialists such as Rockefeller actively patronized new universities through donations and by deploying scientific discoveries and inventions to create economic value. University professors and researchers began to displace theologians as public authorities and advisors on government policy. The universities trained an ever larger number of trained graduates who applied their expertise to economic, industrial, and social questions, helping to initiate the knowledge economy. In this way, the university structure and its nexus, with government, industry, and the public, was formed. This novel combination catapulted the United States to front-runner status in world research, surpassing the Germans and the Chinese.

How well did India follow the scientific revolution in higher education? Our universities are far behind. India has one institution, the

IISc, among the world's top 400 universities. Table 9.2 compares some top world universities with some key Indian institutions.

Indian universities are much smaller in scale than other world universities. Their annual expenditure is 15–20 times smaller than world-class universities in other countries.[5] Our research expenditure is 8–10 times less than MIT and Stanford. The number of faculty for US universities is 3–5 times ours and around 6 times for China. The number of high-performing faculty for our universities is in small two digits. US universities have 1.5–2 times our number of PhD students, and China has it 3–5 times. This is after a surge in PhD students in India during the last 10 years. Similarly, our patent applications and technology transfer revenue is lower by an order or more.

Our universities do not have the scale and critical mass to lead in research. To become world class, they must have multiple disciplines, a large number of quality faculty, and a talented PhD cohort. What else is missing? We will discuss three key elements that are in short supply: autonomy and leadership, fund-raising, and marketing.

Autonomy, Leadership, and Accountability

Research universities must be allowed to make their own decisions. As with any high-performing, cutting-edge enterprise, they must enjoy sufficient autonomy to take their own academic, financial, and administrative decisions. Attempting to control their operations by government is as conducive to success as was our previous experiment in centralized economic planning (Five-Year Plans).

India's top institutions, like those of other countries, are public institutions funded primarily by the government.[6] They enjoy reasonable autonomy over academic matters, but their administrative and financial activities are governed by regulation. The IITs fall under the IIT Act of 1961.[7] Each IIT has a governing board, while the collective IITs are governed by a council of IITs dominated by government officials. The council acts like an interface between the IITs and the government. The council in turn appoints the members of the

Table 9.2: Comparison of Indian Universities with Some Top Global Universities

	IISc	IIT Mumbai	IIT Madras	MIT	Stanford	UC Berkeley	Tsinghua	Zhejiang University
Annual expenditure (in $)	452 M	388 M	167 M	3 B	5.1 B	2.68 B	3.34 B	2.95 B
Sponsored research (in $) expenditure	121 M	143 M	118 M	1.47 B	957.2 M	230 M	–	–
PCT/US patent applications	6	16	13	341	655	263	500	–
Technology transfer/royalty income (in $)	0.12 M	2.4 M	1.07 M	62 M	108.6 M	15.1 M	–	–
Faculty	431	585	558	1,872	1,615	2,258	3,395	3,601
PhD students	2,633	2,471	2,471	3,878	–	5,565	11,903	8,931
Master students	934	5,866**	2,000	2,818	9,771*	5,140	18,661	14,142
Undergraduate students	504	4,004	~4,000	9,880	6,999	27,500	15,636	~23,897
High-performing faculty****	63	81	66	534	642	942	721	–

Source: All figures are from 2015–2016 as per availability from annual reports or the institutions' communications office.

Note: *Dollar conversion is based on PPP. Blank cells represent the unavailability of relevant data.

** These numbers include both Masters and PhD students.

*** High-performing faculty is calculated by multiplying the total faculty number by the proportion of researchers with more than eight citations (as per Microsoft Academic Research data; for more details, refer Chapter 4).

****M stands for million and B for billion.

boards of the individual IITs, including the directors of each institution. The president of India appoints the chairperson of the IIT board. Through these mechanisms, the government exerts great influence over the board composition and choice of the institute leadership. The appointment decisions are usually nontransparent, fraught with delays, and vulnerable to political influence. By contrast, university-governing boards in the West typically elect their own chair and institute president, whether they are public or private institutions.[8] Public universities have nominated government officials.

By the stipulations of the IIT Act, the individual boards do not have the power to make financial decisions.[9] Each year, the IIT council allocates the operating budget for each school with funds from government nonplan grants. There is no transparent and merito-cratic mechanism by which a high-performing ambitious institution can solicit more government funds for supplemental projects or expansion. The government decides how they may spend their money to expand and grow. The IITs also do not decide compensation, promotion, incentive plans, or qualification criteria for their people. The criteria for hiring and promoting faculty as well as the salary levels for various faculty levels are controlled by the IIT council and the government. No institution may alter these, whether for market conditions, to tailor a unique mission plan, or to suit their individual philosophy of incentive structures.

The government exerts further influence over the IITs in ways beyond the IIT Act. One oft-seen manifestation of interference is in the hiring, promotion, demotion, and removal of particular faculty and students. The ministry has been known to issue directives regarding cafeteria food preparation, the admission system, and the rules by which alumni may donate to their alma maters.[10] Without delving into the merits of each decision, one wonders why the ministry bothers themselves with such things. Academics and administrators within the institutions are the best authorities on matters such as these. Government interference in these internal operations not only leads to suboptimal results, but it also undermines the confidence of institutional authorities.

One unexpected consequence of the lack of autonomy afforded to institutions is a lack of accountability on their part. As the government decides financial and administrative issues, they bear part responsibility for the institution's success or failure. The institutional administrators have limited ownership, accountability, and incentive to drive up performance. The leader of a university is best placed to manage the affairs of research and grow the school into a world-class university (see "Leading a Research University"). In the present scenario, however, they cannot be held accountable for the performance of their institution because they do not have a free hand.

Leading a Research University

Now, here, you see, it takes all the running you can do, to keep in the same place. If you want to get somewhere else, you must run at least twice as fast as that!

—Red Queen, *Alice in Wonderland*

Great research universities need great leaders. University leaders need to constantly think about how to sustain and grow research leadership in areas of excellence and build leadership in new areas. Funds and resources, public image, and productive partnerships are all essential concerns at which a truly capable school leader must excel.

In academia, who decides which areas to grow or build? Here is where higher education and private industry diverge. Academia is marked by an interesting iterative bottom-up, top-down process with distributed decision-making. Faculty, whose projects meet with big success, get even greater funding from their sponsors to grow their programs. At this point, they demand more from the university itself. Impressed with the researcher's success and seeing this as an opportunity to take a leadership in the research area, the institution showers him or her with rewards: supplemental funding, better lab space, choice of students, more fellowships, etc. If the research win is significant and with large scope, the leadership may decide to hire additional faculty to grow the area. This is how excellence in new areas builds up for a university.

Similarly, the university leadership and department leaders consult when deciding on new department formation. They look to internal champions, star researchers already in their ranks, or area leaders from outside. They try to identify dynamic individuals who are not only successful researchers, but who can attract other similarly remarkable people. And of course, they consider the person's fund-raising record. Without such a champion, a top-down decision from the university leadership would fall flat. As some of the new department see big success, they rise in standing within the institution, which allocates more resources allowing them to expand. The growth of a research university is all about running experiments and "accumulation of success."

In order for research to grow, other things are also needed: new interdisciplinary centers, new study programs, and collaborations. For example, research in multidisciplinary fields such as data science and neuroscience can greatly benefit by centers where there is shared equipment, funds, and events at which faculty can meet and embark on new collaborations. New fields in particular require study programs to supply a cohort of trained PhD students to enable research success. Collaborations between institutions, such as an engineering school collaborating with a medical school, partnerships with corporations, government agencies, and even across national borders when universities establish extension office, all help grow research. Each collaboration involves the exchange of resources and expertise, new application areas, and assistance in transforming research results into public goods.

The university leadership is vested with the responsibility of acquiring resources and distributing them strategically and productively. However, distribution is a shared decision, made in partnership with stakeholders throughout the university community. To perform effectively, the person in the top job must understand the process of research deeply. His or her actions can make or break a university.

A secondary effect of this government micromanagement is the dearth of professional competence in the higher education administration field. There are hardly any people in India who would know how to run a high-performing university, establish an institutional mission, and manage/grow the school's resources if they were given the power to do so. Government control does not lead to good academic outcomes. The ministry does not employ academic people, nor are their people familiar with the day-to-day running of education institutions. So, how can they be expected to take good decisions?

Given that the institutions are financed by taxpayer money, is this level of government oversight fair?[11] In other countries, the best research universities are also public, but they enjoy significant autonomy nevertheless. For example, both UIUC and UCB are ranked among the top 10 research schools in the United States. They are both public institutions and in addition to their regular operating budgets, they receive more public money (including federal) in the form of research grants. Still, they are governed by independent boards that decide university policy. The boards do seat a representative from the state, but in no way does this entail government control. The boards determine internal personnel decisions such as compensation, hiring, and promotion, or they may even delegate these matters to the departments! This autonomy also implies accountability. The government representative, or the media, might question why they gave an unusually high salary to a particular person, and the board will have to defend its actions. Such a case occurred recently at UCSD when the board awarded a high salary to incoming chancellor Pradeep Khosla. Despite the criticism, the board stood by its decision.[12]

If our institutions are to become world class, they must enjoy greater autonomy. With autonomy, accountability and competition will follow. Government involvement in university decision-making, both by legal and informal mechanisms, inhibits normal incentives for achievement.

Raising Funds to Grow

December 31, 2016, e-mail in my inbox:

Mr Aggarwal, today is your last chance to
support MIT in 2016.

Our community has been working to solve the world's most
pressing challenges, and we want you to join us on our mission.
If you're excited about where we are today and where MIT is going tomorrow....
Make a gift before midnight.

It only takes a few moments to make a positive impact.
We ask that you give $25, but any amount will open doors
of opportunity in the coming semester.

Research activities are a cost center. The investments can be huge, but the activity itself does not directly produce a product that can pay off the amount that was spent on the faculty, students, labs, and infrastructure necessary to perform the research. The goal of research is not to produce a saleable item or service, but to create new knowledge, for example, a fast algorithm to detect a prime number. That may or may not be deployed in the future to inform the development of new products and financial return. Research entails many experiments, many failures, and many incremental steps and results in few of the types of disruptions that actually create long-term value.

The research institution does not necessarily reap any part of the value that gets created from research. Rather, the research enters the public domain as a public good. Taxpayers accrue the return on their investment when the research trickles down into the economy to support private sector companies and entrepreneurs, creating new employment opportunities and contributing to material well-being, medical advancement, and public health.

Research universities have some direct sources of revenue. The most direct is licensing fees from patents based on new research results. Another source is consulting fees paid by industry for researchers' expertise. In both cases, the faculty researcher earns the money, but the institution takes a share. However, these advantages come by way of applied research with industry relevance. The model does not work for basic research or for those temporally not in vogue in the business market. Tuition paid by students comprises a university's next major revenue source.[13]

These revenue streams combined cannot support the total expenses of a university—least of all the $3 billion annual operating budget of MIT. Universities largely depend on money from the government and donations, and gifts of various kinds. The different revenue streams are reserved for different functions. Operating budgets come from annual income such as tuition and student fees, endowment income, and state funding (if the school is public). This money pays for salaries, rents, fellowships, campus upkeep, etc. Public institutions like UIUC use government funds and also have a corpus of $3.3 billion. The money for research and "hiring" PhD students comes majorly from government sources and in part from industry sponsorship. A separate budget funds expansion, such as new infrastructure, new research centers, and additional faculty and graduate student positions. This money is usually donated by alumni, private individuals, and private foundations. Bill and Melinda Gates donated money to build the CS & AI Lab at MIT; one tower is named Gates Tower.

Fund-raising is a major function for any top research university. It is a primary responsibility of any senior administrator of a top research school in the United States. Henry Rosovsky puts it neatly for university leaders in the United States: "Cultivate the art of asking people for money; your career may depend on the results."[14] The senior administrators are assisted by a professionally staffed fund-raising office. MIT has a Resource Development Office that consists of 12 teams, handling various aspects of raising money.

I have firsthand experience with their efforts. In addition to their periodic e-mails, I was visited by a team from the MIT alumni office in 2015, in India. We met at a hotel in Delhi, where after a 30-minute discussion of various things, they handed me a nicely printed brochure detailing their alumni-giving program. They requested a gift,

Table 9.3: Annual Budget, Endowments, and Private Sponsored Research Budgets for Universities

	IIT Madras	MIT	Stanford	UC Berkeley
Annual expenditure (in $)	$167 M	$3 B	$5.1 B	$2.68 B
Endowment/corpus (in $)	$110 M	$13.18 B	$22.4 B	$77.9 M
Private sponsored research (in $)	$11.4 M	$128 M	$300 M	$80 M

Source: Annual reports or institutions' communications offices.

Note: M stands for million and B for billion.

mentioning a moderate number, indicating that a smaller gift would be appreciated too. They explained that I could donate directly to a specific program and also explained the advantages of unrestricted giving. We had a pleasant chat. The beauty is that they would spend time with me when they had so many alumni millionaires and billionaires who could provide greater return for their efforts. Actually they truly appreciate every gift, whether $25 or seven figures. Every dollar counts.

In India, the major research universities are funded by the government. Our annual budgets are 15–20 times smaller than those of world-class universities in other countries (see Table 9.3). IIT Madras has an endowment of around 190 crores, which is 10 times smaller than MIT and 20 times smaller than Stanford (in PPP terms).[15] Sponsored research funding is 7–10 times smaller.

Our institutions are slowly acclimating to the idea of private fund-raising.[16] IIT Madras has private sponsored research of around 19.5 crores (2015–2016), which is 7–25 times smaller than world-class institutions. IIT Madras has set an endowment target of $100 million by 2020 and has constructed a website to present the various reasons why one should donate. Another IIT also asks for donations on their website, but they cite no reasons why anyone should want to donate. They just present you with an amount and options for payment.

Our institutions have a long way to go in their efforts if they hope to realize significant revenue from fund-raising. It cannot be done ad hoc. They need a professionally staffed fund-raising office with active participation by university leadership to develop and manage fund-raising campaigns. This activity is like any sales task, requiring marketing, events, cold contacts, private meetings and visits, and successful closures. While the professional staff handles the groundwork and administrative functions, the university leaders must lend their prestige at high-profile events to appeal directly to wealthy potential donors. They cannot simply rely on passive webpages and alumni groups working pro bono.

Second, executing a successful fund-raising campaign depends on a healthy culture of fund-raising. The United States is well suited to this task, with its ultra-capitalistic society and its unique university structure wherein every person understands the constant need for additional funds so that their professional lives can go on. As mentioned earlier, each faculty member must raise her own money if she wishes to conduct research, travel, support her PhD students, and receive summer salary. Nothing can be taken for granted. However, this mind-set does not always sit well with some individuals, who may believe that the creation of knowledge is a pure activity that should not be discussed in the same breath as money. One of my favorite Indian scientists, Jagadish Chandra Bose, refused to patent his inventions because he thought his research was above commercial purpose, and he did not want to fall into the trap of "greed" (see "Bose's Letter to Tagore"). But he later underwent a change of heart and decided to patent his inventions—ironically, after persuasion of the monks, Vivekananda and Nivedita. Bose holds the world's first patent on solid-state devices, which is the foundation of modern electronic chips.[17] Obsession with fund-raising carries its own ills (as we sometimes see in US institutions) but failure to appreciate the role of economics in research just does not work.

Bose's Letter to Tagore

Jagadish Chandra Bose wrote the following letter to Rabindranath Tagore in reference to his invention of receiving device for wireless.[18]

A short time before my lecture, a multimillionaire proprietor of a very famous telegraph company telegraphed me with an urgent request to meet me. I replied that I had no time. In response, he said that he is coming to meet me in person, and within a short time, he himself arrived with patent forms in hand. He made an earnest request to me not to divulge all valuable research results in today's lecture: "There is money in it—let me take out patent for you. You do not know what money you are throwing away," etc. Of course, "I will only take half share in the profit—I will finance it," etc. This multimillionaire has come to me like a beggar for making some more profits. Friend, you would have seen the greed and hankering after money in this country—money, money—what a terrible all pervasive greed! If I once get sucked into this terrible trap, there won't be any escape! See the research that I have been dedicated to doing is above commercial profits. I am getting older—I am not getting enough time to do what I had set out to do—I refused him.

The letter shows Bose's nobility of thought. However, one may question, what good things Bose could do with the funds, had he accepted the offer.

One must consider the practicality of the situation, beyond the idealistic arguments either for or against the act of raising money. The pursuit of funds serves as a check on the merits of one's work. Soliciting funds allows you to gauge the public's perception of your work and hear criticism. It can be a mechanism for self-improvement. A researcher's or institution's fund-raising ability is another metric in the value of its work, alongside publications, citations, and building systems. As we deal with new economic realities including various crises and the tightening of the government purse, the ability to fundraise becomes a strong argument to justify continued funding from more traditional sources.

Third is to do with the confidence to ask for funds. Henry Rosovsky relates a story of a meal with an alumnus at which he and the Harvard president demanded $10 million from an alumni who offered only five. This is chutzpah, and it works! They got the money.[19] Many administrators do not have such audacity, especially to ask big checks. However, our problem is somewhat different. Our alumni share the lack of confidence in university administrators and are skeptical that their donations would be put to good use. Having attended these institutions, many view them as mediocre in research and overly bureaucratic. Some Indian philanthropists such as Ratan Tata and Murthy have donated to Indian schools, but they contribute quite a bit to Western institutions such as MIT and Harvard. They believe that Western universities are better able to utilize their money to realize their visions for scientific advancement, delivering the best bang for their bucks.

Our universities can change this perception by putting forth their core competencies. They need to showcase the impact achieved by research from funds donated in the past and what can be achieved by the funds of a potential donor. Such objectivity is needed in order to inspire confidence in donors and donor entities. They may welcome targeted giving specifically: mission and program-based grants rather than the traditional unrestricted giving. We already have an advantage over the West in producing lower cost research. If we can put our act together on better research, clearer outcomes, and good marketing, we can make a compelling case to potential contributors. We need to go from a position of hopelessness to chutzpah!

Finally, the government must allow institutions the freedom to raise unrestricted funds and use the funds as they see fit. In 2003, the Ministry of Human Resources sparked a major controversy when they instructed people to donate directly to the government rather than to specific public institutions. This counterintuitive move was soon shelved. Another time, the government placed a cap on the amount of money that our public institutions were allowed to raise. Our institutions should enjoy wide autonomy in an area so central to their organizational health and survival. And rather than restricting fund-raising efforts in higher education, the government should be encouraging our universities to take more responsibility for their finances. For example, they could provide matching grants based on the sums raised privately. A great idea would be to allow industry to redirect their corporate social responsibility (CSR) fund payments to the public universities and to the programs of their choice.

In summary, our universities must learn to embrace fund-raising as a core activity. I say without any exaggeration that private fund-raising will become a necessity in the new world.

Thought Leadership, Marketing, and Outreach

But for the nation's long-term security, prosperity, competitiveness and health, and for generations of lasting new jobs, we must also rebuild another kind of infrastructure now eroding—by renewing our national commitment to fundamental science.... Scientists are like entrepreneurs: They have an eye for spotting unrealized opportunities. It can be hard to predict where those leads will take them. But as we have seen over decades, basic science leads to the new knowledge that leads all of us to the future, along the way spinning off powerful new tools and educating a new generation of scientific pioneers.

—Rafael Reif, President MIT, *Wall Street Journal*,
Op-ed, December 5, 2016

Universities have a social responsibility to advertise and promote their work not only to the research community but to the larger world beyond. Society accrues the benefits of research when new knowledge "spills over" into the public and private sectors, leading to social and economic value. For instance, industry learns about new algorithms of machine learning and applies them in their work to build better image recognition and recommendations systems. Medical professionals learn more about the new research on causes of cancer and recommend precautions. Government could herald a revolution by recognizing the public health benefits of new

efficient diagnoses machines and deploying them at scale. Also, with
the spread of cacophony and disinformation in ubiquitous media,
universities and researchers can add a voice of sanity and objectivity
by subjecting various claims to scientific evidence.

In the last century, a tremendous number of questions of social
and business importance were answered best by science. Today,
public intellectuals take up their cause. Thomas Piketty has enlight-
ened us on inequality in the world through his book *Capital*. C. K.
Prahalad has told us about creating services and products for
poor people. Siddhartha Mukherjee has told us the story of cancer.
Erik Brynjolfsson and Andrew McAfee have described the new
machine age. Others contribute through articles in newspapers
and commentary on news stories. Without such people, many of
them university professors, the world would just slip in to darkness.
The work of researchers and other intellectuals are assets to society,
things of great value. The universities must take care to translate
these works into the language of the public and disseminate them
far and wide.

Effective promotion also testifies to the value of education,
research, and the university's place in society. Given that universities
are cost centers rather than revenue streams, they need to explain to
their various constituencies the value that is derived from the public
investment. All stakeholders need to understand and be reminded
why research is useful, why they deserve funding, and why the brightest minds should want to join them. Sound marketing aligns public
opinion in support of both public and private funding. One Indian
bureaucrat remarked to me that the government has a weak ear
for scientists and science. Part of the problem here is the lack of
"packaging" to make scientific work comprehensible and sellable.

India does not show much deference to scientists, and it seems
we do not look to science to guide our way. Indian public discourse
on social issues is infested with politicians, activists, and generic
celebrities. The idea of specialists has not taken ground. A host of
showman intellectuals expounds opinions on anything and everything, and the media laps it up. One seldom sees an Indian university

professor on a news channel, commenting on a story. Two of the biggest mainstream newspapers in India hardly have any coverage of Indian science news. It is appalling to see how they devote almost a page to sensational scientific news, almost always from the West. On the other hand, some newspapers such as *The Hindu* and *The Indian Express* do report on some domestic scientific developments. Consider China by comparison. On a trip to Beijing in 2016, I was pleasantly surprised to see science coverage on page 4 of *China Daily*, China's preeminent English newspaper. One story was on China's space exploration program, and another featured a controversial paper in *Nature* related to CRISPR[20] by a Chinese scientist. A third story described an experiment in random number generation by a Chinese graduate student at CERN.

The top Indian research universities conduct fledgling outreach programs. IISc runs a Science Media Center that publishes stories on their page and also distributes stories to the mass media. There are two members on the team. The website lists around 30 contributions to mass media publications in three months (October–December 2016). This is a good start. The websites of the IITs have sections on research news and research highlights, but they are generally not properly maintained: the stories are not updated, are poorly written, and the sites are pervaded by nonresearch-related content.[21]

Let us take one of the IIT homepages as a case study. The top key space is reserved for a research highlight. For January 2017, I see a paper title that would be incomprehensible to an educated audience. The space features an image, part of which is pixelated. Upon clicking the link, we see the paper abstract and links to access the full paper. This page does not feature any links to other research highlights. The left menu bar has a link called "Research Highlights," but upon clicking it, you arrive back at the current page. The next few links on the Research@IITx page are repeats and not relevant. This is followed by a list of a few papers or news briefs for each school department, which are not dated. For a university that aspires to world ranking, what better word for this attempt at outreach than "terrible!"

Western universities take outreach very seriously. Recently, the president of MIT published an op-ed in the *Wall Street Journal* arguing against funding cuts for science research. This is an example of outreach at the highest levels—from the president of one of the world's preeminent institutions to the president of the United States—and a testament to how seriously education leaders consider their mission and the importance of outreach in furthering their mission. The MIT News Office, part of the Office of Communications, has 18 members and actively disseminates news about MIT's research to the world. They have professional science writers who specialize in translating scientific developments for a general audience. The school's website features a new research story every day. In addition, the school operates the MIT Press, a publisher of research-led books and *MIT Technology Review*, a general interest magazine for science and technology. These resources inform not only about MIT but also about interesting and groundbreaking research wherever it is happening.

Even the professors and graduate students at the university have a great understanding and interest in marketing. People often quip that success is 20 percent good work and 80 percent packaging. However ridiculous it may seem, people devote significant amount of time to building neat presentations that tell a comprehensive and compelling story to communicate the big-picture impact of their research work. They spend considerable time rehearsing talks and giving mock presentations to colleagues. Aware of the importance of attention in generating funds, they generally premarket a vision, celebrating a success even before it happens in order to align the resources. One such example was the "one laptop per child project." The idea was to give a $100 laptop to every child in the world. The project didn't suggest a proper plan on how kids will use the laptop, how it will help their development and also, who will pay for the laptops. The project was highly publicized and covered by all major publications. It finally resulted in a failure. Such presumptuousness and brazen marketing would frustrate many, but it serves a good purpose nonetheless. Many of these experiments fail, but the few that succeed justify the time and resources spent.

Our top universities need to establish professionally run marketing and outreach departments. Indian universities face a credibility challenge. It is not enough for our faculty to increase their research ambitions; they must also become good marketers of their work. Academics and institutions must leave their ivory towers—it is their responsibility to simplify their work for everyone's understanding. They are public institutions and need to connect with people. In today's world, it is essential.

The Research Funding Ecosystem

Among G20 countries, India ranks ninth in money spent on research (in PPP). The United States and China top the list; South Korea, with a GDP one-third that of India, spends 68 percent more. As a percentage of GDP, our research spending ranks 15th. We cannot attribute these figures solely to the low rate of investment by private industry; in terms of government expenditure on research as a percentage of GDP, we rank 13th. Overall, we spend less than 1 percent of our GDP on research. By comparison, the United States spends close to 3 percent, and China 2 percent. The United States spends as much as 14 times as much as India on research, while China outpaces us by 5.5. Our research spending in absolute terms and as a percentage of GDP is much lower than the countries we compete with.[22]

Several government agencies in India provide funding for research. The most prominent include the DST, DBT, Department of Atomic Energy (DAE), CSIR, and the Science and Engineering Research Board (SERB). The DST and SERB provide funding for science- and engineering-related projects. DBT focuses specifically on projects in biotechnology, while DAE sponsors research in atomic energy and material science, among others. DST has an annual budget of around ₹4,000 crore, DBT of ₹2,000 crore, and SERB around 800 crore. Researchers from public and private institutions can submit proposals for research and request funding. Committees of experts review the proposals and then recommend which are deserving of sponsorship.

In India, there is a wall between university research and government research. University researchers are not typically involved in mission-driven research, whether space, defense, or other areas of focus. Our space research organization (ISRO) and defense research organization (DRDO) operate fairly independently with their own set of researchers. ISRO has a program called RESPOND for partnering with academic institutions. It contributes 7.5 crore for 47 new and 41 ongoing projects.[23] DRDO also has a partnership program, but it is fairly small scale.[24] By contrast, NASA's funding to universities amounts to almost half a billion dollars annually ($431 million in 2011)[25] while the US DoD sponsors a couple billion dollars in university research each year.

This separation between defense/space research and university research does not serve Indian science well. On the one side, government programs lose out on the benefits of university research. On the other, Indian universities are denied a source of funding, a supply of problem areas and research questions, and opportunities to work in applied science. By sharing resources and expertise, these government agencies and universities could realize large-scale collaboration that would facilitate the progress of Indian science.

University research funding agencies do not have many research themes or mission-driven programs that invite proposals. DST offers 5–15 of these in areas such as clean energy, nanotechnology, and electrical waste management. By contrast, DARPA (USA) currently has 250 research programs that solicit proposals. DARPA keeps an eye on where the next technological breakthroughs are likely to occur. Program managers, that include academic researchers recruited for short-term assignments, introduce programs and put out requests for proposals in these areas (one recent example from DARPA is "Explainable Artificial Intelligence," the next frontier in AI). This system helps to attract a critical mass of researchers who put forth their best efforts in order to win the available funds. This accumulation of efforts lead to breakthroughs and accelerate research progress.

Agency funding of academic research is beset by problems. In our survey, Indian faculty rate ease of process for funding as the lowest among 11 parameters of academic research quality. Mostly, they complain of bureaucratic interference, delays in the form of excessive paperwork, and cronyism. Two other complaints are also notable. First is lack of feedback. Researchers often complain that they do not get proper feedback on why their application was rejected. Oftentimes they receive a single line stating "no prior experience in the field." Feedback is an important aspect of the research proposal process. As one relies on detailed reviews of paper submissions, one relies on research proposal feedback to understand how to tweak or alter the research project to make it more viable and how to write a more effective grant proposal. Proper feedback also gives the researcher confidence that his or her proposal has been respectfully considered and gives him or her the opportunity to address or clarify any problem areas or shortcomings. A one-line feedback report is a sure way to douse a person's ambition.

The second complaint concerns the reluctance of funding agencies to sponsor big, risky projects. Agencies shy away from large projects that require significant resources. They prefer to spread their money thin, in socialist fashion, to benefit a larger number of researchers and give each person a share of the pie. One researcher notes:

> The canonical reason of rejection is that the project is too ambitious, as if it is a crime to set high ambitions or to challenge current thinking. Sometimes I wonder whom are they fooling. Probably themselves!

DARPA sees their mission differently: "DARPA operates on the principle that generating big rewards requires taking big risks."[26]

The Indian way is counterproductive. To achieve real progress, our agencies need to support high-risk, high-reward projects. After all, if we know something will work (i.e., the project carries little risk of failure) then it is not research. Projects that explore the realm of the unknown will create impact, garner citations, and create economic value. Many of them may fail, but knowledge is gained through failure, and one big success will eclipse the many failures that came before. We should regard research very much like start-ups, where the venture capitalists are eager to fund those for which there is the potential for great rewards.

The agencies have taken note of some of these issues. VijayRaghavan, Secretary of DBT, has acknowledged the slow approval process and attributes the delay to the sheer variety of proposals they get for projects of various size and scope. Under the current system, all proposals whether large or small undergo the same process and take the same amount of time. He has proposed dividing the proposals by category and size of request as a first step toward streamlining the process and reducing bureaucratic hurdles and response time. There has also been progress in digitization and the use of web-enabled submission and review processes, which have been a long-standing demand.[27] The agencies have also begun to show more interest in larger and more risky proposals. While these efforts are welcome, so far they have fallen short of expectations.

Let me end on a positive note. One long-standing problem has been funds for new faculty, who are still learning the ropes and are yet to write their proposals. Institutions have begun to set aside funds specifically for new faculty, generally up to 25 lacs for a period of three years.[28] Although the funds might be slim when the researcher is involved in applied fields requiring him or her to outfit a lab, the program is welcome nonetheless.

10

Leading Science and Technology: Vision for the Future

[T]o make a great future of India, the whole secret lies in organization, accumulation of power, co-ordination of wills.... This is the secret, accumulation of will power, co-ordination, bringing them all, as it were, into one focus.

—Swami Vivekananda

Let us work toward making India a scientific superpower. We can be inventors and discoverers again. We have the experience and the potential. In the present state of technological and economic change, now is the most opportune time to embark on this national journey. We have considered a number of issues with our current research ecosystem in previous chapters. Now we look at how we can improve our situation.

I begin this chapter by discussing certain general guiding principles for forwarding the research agenda. These principles, if followed, would help create an ideal research ecosystem. We can distill these principles into particular objectives for India organized by impact area. These principles are also a guide to build policies and a yardstick—the dos and don'ts—what actually helps, what impedes research, and what actually makes no difference!

The steps to achieve our objectives are many and diverse. Some we can realize in the short term while others will require patience. Some will be evolutionary while others may seem revolutionary. And they will require the participation of multiple stakeholders. All of you could be worthy contributors of ideas—good ideas come from everywhere—this is very much the spirit of research. In fact, this is the very spirit of research!

Sixteen Principles for Building a Highly Effective Research Ecosystem

Will and Ownership

We need a strong belief in the virtues of science and a tenacious will to promote scientific research as a national goal. To embark on such a mission requires a buy in from everyone—starting with the media—the opinion creators, the politicians, the bureaucrats, the institutional leaders, and educators. Each person at every level should be dedicated wholeheartedly to pursuing the research agenda. Ideally, such individuals should have been deeply involved in research at some point in their career. They need to be believers, progressive thinkers, and charismatic go-getters. We need a laser focus on research and people dedicated to the task, at all levels.

Plan for the Future

When considering national policy, economic development, and scientific research, we need to build institutions and plans with a 10–20 year focus. It does not suffice merely to be reactive to just our immediate problems nor is it productive to attempt to harness immediate opportunities only. The world is changing too fast. We are witnessing the advent of new technologies and new kinds of businesses that present us with new problems and opportunities. We need to consider the big picture: predict, plan, and revise continuously. Second, we need to save ourselves from our habit of applying only "Band-Aid" solutions that are neither robust nor sustainable.

Last, we need to take into account the time it takes to implement anything systematically, specifically when being executed by the government. We should invest in sustainable long-term measures, otherwise they end up being too little too late.

Understand the Centrality of "People"

The quality of research cannot exceed the quality of the researcher. If one wishes to reap the economic and social rewards of great research, one must attract the best minds to scientific careers. Talented, capable people simply need the right incentives and a professional environment. As caretakers and practitioners of high knowledge, they must feel respected in the society and their institutions in the same way that knowledge itself is respected. Without the right people, no amount of money, infrastructure, or policy can produce transformational results.

Let Merit Be the Key Principle

As a moral principle, resources for human development should be equitably distributed. But in pursuit of this moral principle, allocation of our resources for research should be merit-based and competitive. We must allocate resources to those individuals who demonstrate that they have the best research ideas, who write the best proposals, and produce the most promising results. We must be open to many different ideas and thought processes, but nothing mediocre. We should put power behind our star performers with the money, facilities, equipment, student support, and decision-making power that they need to further their success. And we need to celebrate their success and that of their institutions so that they may inspire others to excel.

Promote Autonomy and Accountability

Researchers and research institutions work best when they are independent. No one should "tell" them what to do. They should be driven only by their own mission and curiosity. They should have the power to implement their plans independently. Bureaucratic rules, regulations, and micromanaging hinders creativity and progress. At the same time, they should be held accountable for their results[1] and ethical practices. Resources should be allocated accordingly.

Create Differentiators

To compete globally, we need to reinforce our current strengths while simultaneously building new strengths. Our current strengths include our proximity to certain problems, our ability to run certain experiments that might be more difficult elsewhere, and the availability of higher performers in certain areas and cost/skill advantages. We can build new strengths by developing new innovative institutional structures, identifying promising but neglected areas, and developing new research processes and policy. By copying others, we can play catchup. In order to lead, we need to prepare for ingenuity.

Create Areas of Excellence

We cannot excel in everything. A careful consideration of our national priorities, research velocity, and particular strengths will help us identify the best areas in which to focus our initial efforts. Having identified the right areas, we need to accumulate efforts in these areas—having the best researchers, the right infrastructure, and seeking to solve the toughest problems. Disruptive research happens by accumulation and multiplication of success of several researchers in a healthy ecosystem. Mission-based programs and institutions is one good way of doing this.

Raise the Bar

Top papers and citations are great, but they cannot be goals in themselves. We must raise the bar and aspire to solving our biggest puzzles and toughest problems. We cannot be mediocre in our choice of problems. We can expect to solve some problems, make progress with others, and sometimes stall. For these, we should not bring down the dream. Real breakthroughs come about by dreaming—it is a culture and a way of working with perseverance and grit. At the end, we need to benchmark ourselves on the big ideas and solutions that come out of India.

Respect Ideas, Respect Failures

We need to respect ideas: not just small, safe ones, but big and crazy ones! Science cannot afford to be complacent and arrogant. Science keeps surprising itself over the ages.[2] We need to be open to the potential of new, strange ideas and provide resources to test them. When we support radical ideas, there will be failures. We must accept that this is so and remember that failures also impart knowledge. The few successes that we achieve will outweigh a slew of failures, and our mistakes will help us learn.

Create Awareness

No one can be forced to do research. Good research never results from compulsion. The way to enable research is to create awareness about it among all—the government, our institutions, the industry, and the industry and the society, at large. Awareness of research will celebrate its virtues and provide role models. It also is about the larger virtue of scientific thinking itself. By such awareness, the intellectually curious will steer themselves toward research automatically. These self-selected individuals will become the best researchers.

Create Competition

Top performers can be spurred to even better performance through competition. Competition is a key factor in motivation, pushing people and institutions to excel, take risks, and explore the out of the ordinary. There are many examples of competition inspiring people to move out of their comfort zones, try something disruptive, and thereby achieve success. We should create challenges for researchers to compete. We should distribute research resources and funds competitively. The process needs to be world class, transparent, and fast.

Remember, Money Is Not a Bad Word

Money holds people accountable, provides incentives, and helps align interests. We can use money to encourage all of these things. With money as a reward, people have a higher potential for performing and a higher chance of delivering. The availability of money leads people to collaborate in pursuit of it. Money has the potential to create knowledge. It is a resource, not an end. It should be used for its very important role.

Create Incentives

The enigmatic curiosity and virtue of the researcher is not enough to support research. Individuals and institutions perform better under the right incentives, which must be professional as well as monetary. A researcher cares most for an environment that helps multiply their success and makes big impact. Such things are at least as important to success as a large paycheck. As a nation and our institutions needs to make these possible—these are the greatest incentive for researchers.

Look for Alternatives While Working on Foundations

There are new ways to research, raise money, and disseminate research results. Crowdsourcing, hackathons, *jugaad*, 3-D printing, and social media are some examples. We should delve into these and use them to advance knowledge in the best way. However, nothing replaces well-functioning universities, PhD programs, transparent and merit-driven funding agencies, and communication and collaboration in the form of conferences and journals. These are all essential components of the research ecosystem. New innovations will never diminish their necessity. They shouldn't be an excuse to neglect these.

Make the Market

Research is a public good. Around the world, governments are the largest funders of research. The beneficiaries of this funding are private players and individuals, who convert their research into products and services that improve our lives and provide economic gains. The government's role as sponsor entails making a market for private actors to participate in all parts of the research ecosystem. Government measures could include tax breaks for conducting research and for research philanthropy, colocating research institutions and academia, IP help, and public funding of innovative companies deemed too risky by private lenders. This is useful for scaling research, but also essential for deriving tangible outcomes.

Be Dispassionate and Transparent

Good science is beyond social, political, or religious considerations. Research must be approached dispassionately. The researcher must be prepared to accept the results, regardless of the degree to which they conflict with prior assumptions and hypotheses. A researcher also must have the courage to reveal the results without fear of political and social implications. Political actors should not seek to silence or censor public intellectuals, regardless of whether they are public employees. Scientific voices must be independent. Noam Chomsky, Richard Stallman, Thomas Piketty, and Thomas Nagel, all professors at top universities, have held controversial views on political, scientific, and ethical issues. They continue to enjoy the support of their universities. Donor agencies and universities should vigorously support the autonomy of researchers and the research process and never seek to bend the proposal to serve an ulterior goal.

What India Needs to Do?

Let us move from the abstract to the concrete. What can India do to invigorate our research ecosystem? The government has taken several measures in the last few years.[3] However, most of these have been "Band-Aid" solutions to fix bad policy rather than an attempt at structural change. What structural and policy changes would be actually helpful, at a national and institutional level? Let us discuss these now.

Focus on Universities as Places for Research

In India, four times as many researchers work in government research labs as in university labs. This contrasts poorly with countries that demonstrate the highest research output (refer Chapter 9). The world discovered long ago that productivity increases when faculty and students, together, take the lead in research work. The university-based research model creates high-quality research training and a continuous flow of new researchers.

The Indian government needs to promotes universities as the place to do research. A larger part of the current research budget should go to universities. A thorough audit of government laboratories will help identify where funding can be reduced and redirected. Labs that perform well can be retained and aided in growth. Those that are underperforming or engaged in obsolete areas can be phased out in due course. The government had good success in reforming the PSU sector. They have been reformed, prodded to become profitable, and some were divested or even shut down. A similar campaign can help our research sector as well.

We must also encourage university/government partnerships. There is much potential for government labs with strategic goals such as DRDO and ISRO to interact with university personnel and embark on collaborative research projects for mutual benefit. We see models for such collaboration in DARPA and NASA, which offer funding in various problem areas through calls for proposals.

Make Institutions Autonomous, Independent, and Accountable

Government can also be helpful by allowing our research institutions greater freedom to make their own decisions. Government should not dictate how they should spend their money, how much to pay their personnel, how to manage promotions, and how or when to scale up. The research institutions and academics who are directly impacted by these decisions are best able to decide them. Furthermore, political interference on appointments and decision-making, besides being unhealthy and inefficient, goes beyond even the current ruling public charter. Such interference and micromanagement is demoralizing and deters the best people from entering the university system to pursue academic careers. This must stop.

With autonomy comes accountability. Our universities must justify their public funding against clear performance metrics. Public universities should be required to compete with each other for resource allocation. Such an approach to funding will help them improve themselves.

Autonomy begins with the governing board. The boards of public universities should be independent, choose their own chairperson, and decide the leader of the universities. Given that the universities are supported by public funds, government representation is proper. However, this representation should be a limited number of members, including both the central and state governments. The board should be primarily concerned with the institution's progress. They should be free to deliberate independently, balancing the views of all interested parties including the government, to steer the university along the correct path.

Consider a recent example from the University of California, San Diego. In 2012, the board appointed Pradeep Khosla as chancellor with a base salary incrementally larger than the salary paid to the previous chancellor. Gavin Newsom, California's lieutenant governor and the government's representative on the board, was the lone dissenter. However, the board overrode his dissent, citing Khosla's "proven background in fundraising and bringing in research dollars.... Chancellors make a difference, we have to recognize that." Furthermore, the board explained that the additional salary would be mostly covered by foundation money and not by taxpayer money. India's institutions would benefit from such freedom to make decisions without fear of government objection or reprisal.

Second, the boards have power over academic and administrative issues, but not financial issues. The government and the IIT Council makes all decisions on capital investment, salary guidelines, fee structures, and other such matters. Financial autonomy is key to making our institutions world class. Again, autonomy can be supplemented and buttressed by accountability. The government should set certain performance benchmarks in proportion to the funds provided. Accountability should start at the leadership and flow to every faculty member in the institute. The disbursal of future funds must depend on how well these metrics are satisfied. The government has already instituted a similar arrangement with the PSUs, where funds and autonomy are set in proportion to performance.

Overall, university finances should be run in a more businesslike manner, with line items detailing the profit/loss of various university operations. Instead of the present flat grant system, a more sophisticated accounting system will reveal what the government is really subsidizing: is it undergraduate teaching, scaling research, postgraduate programs? When each component is examined individually, the government can deliberate and decide what they want to subsidize and where they want to invest. In turn, this will force the institutions to examine their operations and programs more critically. They may even price their courses and programs differently for the public, creating a market.

These things cannot happen overnight. Until now, our universities have had little experience managing their own finances, making businesslike decisions, and performing research administration. The drive for autonomy will be pushed back by decades if the universities mismanage themselves to the point where the government has to bail them out. Before we can hand over significant autonomy, we need to establish the right management structures and conduct proper training for administrative personnel.[4] In my opinion, we should seek training collaborations with top universities abroad. Training partnerships with institutions in the West and in South Asia will give our administrators a holistic experience. When Hong Kong University of Science and Technology (HKUST) was starting up, the governor insisted that the university head should be president of a top Western university. I do not think that we should set such preconditions for our leaders, but rather select the best person for the job whether in India or abroad.

India liberalized its economy in stages, allowing the market to react harmoniously. We should follow a similar path with our universities: limited autonomy leading eventually to full autonomy. Limited autonomy could mean approving budgets for different line items in the financial statement and adhering to particular overall performance benchmarks rather than analyzing every spending decision. Limited autonomy could also mean more flexible salary bands, rather than the narrow ranges dictated by the pay commission.

The government should treat the liberalization of university system with priority similar to that it gave to economic liberalization. We need a great team to do it. As the government pursues limited autonomy, they must resist the temptation to make it a goal rather than an end. The divestiture of power should be the guiding principle, with full autonomy as the ultimate goal. We need to plan for the future.

Our private universities require more autonomy as well. However, accountability is tricky in their case (see "Autonomy and Accountability at Private Universities").

Autonomy and Accountability at Private Universities

The Indian government also exercises control over private universities and institutions. The government places various restrictions on the programs they can run, the salaries they can award, the number of students they can admit, and the fees they can charge. There are also regulations on infrastructure—the amount of land and property required, and the kind of laboratories and classrooms. These regulations are not always optimal. Different institutions have different strengths and weaknesses, and should enjoy the freedom to make different trade-off choices based on their mission.

Besides, these regulations create space for corruption in regulatory bodies. Regulatory entities have often been accused of bad behavior. It is unclear whether government regulations have actually improved quality or if they have simply created an industry of higher education entrepreneurs who know how to lobby government agencies and bend the rules. These conditions dissuade able and well-meaning individuals from establishing new educational institutions because they do not want to deal with a corrupt system.

We might be tempted to think that we can solve this problem with the same remedy we prescribe for public institutions: autonomy and accountability. However, accountability is tricky in the case of private schools. Ultimately, they are only accountable to the students and parents who are their customers. In an ideal market system, poor performance would lead to loss of customer interest, fewer applicants, and declining revenues.

Education is not a perfect market entity, however. Consumers cannot easily distinguish between good and bad. We do not have objective measurable parameters for comparing institutions. Many universities make up numbers and facts on student placements, number of faculty, faculty qualifications, and collaborations. Others use showmanship and optics that have nothing to do with education or research: marble flooring, chandeliers, celebrity endorsements, and even foreign-accented teachers in advertisements! Consumers are often swayed by such gimmicks—especially the first-generation learners. Therefore, accountability is really difficult.

I believe the solution lies in creating a credible rating system for these educational institutions and widely disseminating the results to the public. The market itself cannot create healthy competition or control the quality of our universities and colleges. Government should not itself run the ratings system, but should create a market of credible rating agencies. The big four consulting companies, education research, and think tanks are natural candidates. The government has already done a similar exercise in credit ratings, establishing approved agencies among which companies and banks may choose their preference.

Once established, the government should mandate that all institutions—public and private—must go through the rating process. They should take it upon themselves to ensure that the school ratings are widely disseminated among the public, through newspaper publication and Internet websites. As the public becomes more aware of which factors indicate institution quality, we can begin to remove direct government oversight and regulation. The best processes for achieving this, and the measurements of effectiveness, would constitute a research project in itself!

Identify Areas for Excellence

No nation, institution, or person can do well in all research areas. Each entity operates according to their talent and comparative advantage. India can share in this world by identifying research areas in which to excel. Three principles can help identify these areas. First, we need only consider our natural strengths. Because we already have a large IT industry ecosystem, computer science is a natural point of concentration. In fact, computer science should be a research priority if we wish to preserve our edge—the dulling of which has already begun to show. Additionally, our demonstrated success in space research raises the opportunity to specialize in world-class instrumentation and equipment produced at a lower cost.

Second, we need to invest in new and emerging fields that show promise and potential to drive the next generation of innovation. Data science and AI are two such fields that have recently seen disruption. Given our IT industry combined with the aptitude of our citizens for computers and numbers, India could share in these fields. I also find that India has an aptitude for the quick and economical collection of data. Could we create large data sets for important problems, be uniquely positioned to solve them, and have all top researchers in the world work with us, our data set? Other important areas are biotechnology and neuroscience, where we do have a fledgling industry and ecosystem.

Third, we should identify areas for specialization by examining those areas that are of our own national interest. Defense research comes first to mind, as manifest in nuclear, mechanical, robotics, material, cybersecurity, and also data science. Another national interest is manufacturing and manufacturing processes, including allied fields. If we wish to successfully convince the world to "Make in India," we should pursue research in production and manufacturing processes. And most importantly, we need to consider public health and medicine, where India has a long way to go.

But who decides in which areas India should invest? The biggest mistake is to leave it to a single politician, bureaucrat, or even academician. What is hot and not in research changes rapidly. Research progress is nonlinear: suddenly a process, technique, or area previously of little importance sees a breakthrough and becomes an area of great attention. For example, the academic community had largely lost interest in neural networks about 20 years ago. Now they have come to dominate AI after Hinton's deep learning work. It has suddenly become the largest focus of investment in the field.

No single person knows all our strengths: they are many and decentralized. The government rightly has a voice on what is of national importance. However, there needs to be other voices, and decisions on narrow areas of research need to happen through a decentralized process involving researchers. Grant-making agencies allocate funds in focus areas. Researchers on the other hand have discussions and alert the program managers to promising areas where they could create new programs. As an area of investment becomes successful, we need to accumulate investment and effort in it and grab a leadership position. Similarly, the heads of universities and funding agencies can meet periodically, to discuss priorities and align goals.[5]

The institution's specialties should also be determined through a deliberative process involving the department heads and high performing faculty. This is a bottom-up, top-down process to identify areas of strength and build capacity to take leadership in it.[6]

Accumulate Effort

To build leadership in an area, the most important requirement is a critical mass of researchers working collaboratively. When researchers combine forces, they make faster progress, can adapt to new methods, and can build on each other's successes. Successful and productive research communities guide the direction of research in the field. Impressed by their success, new talented researchers aspire to join them, they build their own forums such as conferences and workshops, and influence spreads to the industry. In this way, MIT, Harvard, and the United States in general have become doyens of new knowledge, amassing research success for a century or more.

One way to accumulate effort in identified areas is through mission-driven programs. Presently, DST runs about a dozen mission-driven programs including those in data science, clean energy, and supercomputing. Their nanotechnology program has been somewhat successful in kick-starting quality research in this area. The number of programs is too small to be consequential. By contrast, DARPA has 100 program managers looking after 250 research programs. They govern the competitive bidding process to accumulate effort in in their field. The United States dedicates considerable funding to all new promising fields, seeking to identify and extract their maximum potential. Some areas see success and draw more funding while others drop off to await some future breakthrough.

India could use more such mission-driven programs in narrower areas and problem statements, which could also be those of national interest. Worthwhile endeavors include cleaning the Ganga, decreasing pollution in cities, or effectively treating Indian strains of viruses. We can determine these areas of attention involving multiple area specialists and program managers, keeping them in steady rotation to ensure a steady stream of fresh ideas. The competitive grant process raises the bar, motivating people to do their best in writing a winning proposal. Such a process helps to accumulate efforts in areas of national interest and also attracts new talent. The further accumulation of funds, resources, and talent will be in the winning areas.

Currently, our defense and space organizations, DRDO and ISRO, respectively, have minimal interaction with universities. They should invite university researchers to help solve their research problems through call for proposals. This can help accelerate university research through supply of research questions, funding, and solution deployment opportunities. On the other hand, it will create local capabilities in the research areas relevant to them. We must note that defense-oriented research has led to some of history's greatest scientific successes, including innovations in air travel and the Internet.

Like nations, institutions also specialize and accumulate effort in those areas. No single institution can be good at everything. Within the fields of electrical engineering and computer science, the Princeton University is great at theory while CMU does very well in robotics, UIUC in building systems, and Berkeley in circuit theory. Our Indian institutions should also seek to determine their natural area of leadership through a collaborative process. Their specialty could be a function of their local strength, such as areas in which they have great researchers/funding already, proximity to certain industry, or problem space. They should accumulate faculty and resources in these areas. Institution leaders should utilize their wide networks to find champions in those fields outside the university and bring those people in.

A great way to accumulate talent and effort in new areas is to create new institutions that focus exclusively in the target field. These could be specialized institutions dedicated to neuroscience, biotechnology, and materials, among others. Google recently established a Google Brain Project, accumulating multiple star researchers in the field of machine learning to lead the effort. Allen Institute for Artificial Intelligence funded by Microsoft cofounder, Paul Allen, is another example. Such dedicated institutions have the potential to provide leadership in the entire field.

We must acknowledge a contradiction in the notion of accumulating effort in a particular area or creating specialized institutions: today, more and more research is multidisciplinary. In practical terms, this means that we need to build mechanisms for specialized institutions to work collaboratively on problems. There are many options to achieve this. For example, most important is that related institutions working in related fields should be situated in geographic proximity. In Delhi, we find IIT, AIIMS, and Delhi University institutions all in a similar locale. In planning new institutions, we need to consider geographic proximity.

Also, calls for proposals for interdisciplinary research will naturally incentivize people to work together. We should have interdisciplinary centers in the vicinity of institutions. They could house seed funding for interdisciplinary projects, equipment, support staff, and host talks and discussions. Champions recruited through a competitive process should run them.

Make Public Science Aware

Much of our efforts to achieve research eminence will depend on public awareness and support. Multiple stakeholders must coordinate to create greater public awareness about science and research.

For starters, the government should raise the profile of scientific research by including it in its strategic priorities and giving it a place in the national discourse. India's scientific achievements should warrant the same attention as our economic growth numbers, IT power, and start-up culture. Attention should not be limited to ISRO, as it is today. A few words from the prime minister each year could inspire many young people to pursue a scientific career. As the government talks about pollution, constructing toilets, and saving electricity, they should also talk about promoting scientific thinking. How they could attach the abstract idea of scientific thinking to tangibles—like Gandhi's charka—is something I will let them innovate upon!

We can also create awareness through our current educational institutions. Our school curriculum and books should properly cover India's historic contribution to science: objectively, not with jingoism. We must go beyond Arabic numerals and Pythagorean theorems to discuss the contributions of our civilization to science. A focus on our heroes and national pride goes a long way in creating role models and instilling the "can do" attitude in youngsters. Such an approach will have the added benefit of spawning new research into the subject, which the government can sponsor through targeted and competitive grants.

Our top institutions should undertake greater outreach efforts to advertise research achievement. They should establish an outreach office staffed by professionals to periodically arrange public talks, campus visits by school students, and even summer camps. Going beyond, they should establish permanent science-related attractions on their campus. For example, MIT has the MIT Museum which chronicles the scientific achievements of MIT affiliates. Why not establish museums at IITs or IISc? Why not go beyond merely copying MIT's ideas and think how an innovative twenty-first century museum can be that would capture the public's attention and bring them to campus? If Akshardham Temple and Kingdom of Dreams can be on the Delhi tourist map for religion and entertainment, can we not think of anything as engaging for our science legacy?

Here is also an opportunity for news and entertainment media. India's English dailies have sections for business, entertainment, and sometimes spirituality, but not for science. Mainstream TV has little any science-related shows—we can only reminisce about the *Turning Point* from 1991, with the charismatic Professor Yash Pal. We do have Discovery and the National Geographic channel, but these feature little Indi-led content.

In a free market, business must run by demand. However, the airwaves are a public good, and I would argue that the media has responsibility for promoting scientific education and research advancement. I would go further and argue that there exists a latent demand among the public for science programming. We need a champion within these media enterprises, an entrepreneur within, to take a risk and highlight it. The government could help here by sponsoring shows or providing subsidies or tax breaks to media businesses that devote a percentage of their programming to science themes. The government needs to play a key role to seed and build a market.

Make Our Institutions Attractive to Researchers

Above all, we must remember that research is only as good as the researcher. Just as a sports victory depends on the sportsman, research productivity and scientific advancement depend on the researcher. We must attract the best, highest achieving individuals to research careers. Presently we lose them either to careers in industry or to research careers abroad. Indian research careers are not attractive to our best students, who are much better rewarded financially by becoming a software engineer in an MNC, IAS officer, or a doctor.

We should wish that India becomes an ideal research destination not only for our own citizens but for the best researchers from around the world. The top institutions in the United States and Europe are home to a diverse array of talented individuals.[7]

Immediately, we must concentrate on engaging our own best students in research. We are blessed with a critical mass of talented individuals who, when properly motivated, could drive the country's research agenda. Following this, we can market ourselves as a viable research destination to the best from the developing world and countries in geographical proximity. Countries in the Indian subcontinent, Southeast Asia, and Africa are natural choices. Not only would such individuals enhance the quality of our existing PhD cohorts, they would act as bridges to business and education opportunities in their home countries, breathing new life into our research ecosystem. Eventually we can expect to see researchers from the United States and Europe traveling here to work in areas where we have developed a strategic advantage and have accumulated success. Everything is possible, so long as we have both the intention and a defined policy to make it happen.

Awareness

We need to package and sell the research career in India. We can spark general interest by disseminating the research achievements of our top institutions through Indian and global media. Then we need to shine a spotlight on the researchers themselves. Our best researchers should be well-known national celebrities just like our star entrepreneurs, CEOs, sportsmen, and actors. Institutions should connect their research authorities to media outlets who are searching for answers to topical issues of today. These will serve as role models and inspire our youngsters to build research careers in India.

They have an additional task in becoming good aggressive door-to-door salespeople. Our top institutions must cease their passive ways, waiting for candidates to apply. They need to visit and sell the position of research faculty to PhD students at the top institutions in the world. They need to highlight their strengths by making presentations, meeting individual PhD students, and convincing them to join us by making them feel "wanted." Similarly, they need to aggressively reach out to the top undergraduate institutions pitching their PhD programs, and also engaging them through talks. One should also consider a larger marketing campaign, like the Indian armed forces do today with a dedicated website, TV advertisements, and YouTube videos.

A large-scale structured internship program could help greatly. Our top institutions could recruit bright undergraduates to participate in research projects, working side-by-side with faculty and PhD students. All top universities run undergraduate research program for their students—MIT runs Undergraduate Research Opportunities Program (UROP),[8] UCB has Undergraduate Research Apprentice Program (URAP), and UIUC has Illinois Scholars Undergraduate Research (ISUR) among others. Our institutions should aim to recruit hundreds of interns both from the institution and outside, at least at the rate of 1.5–2 per research faculty. The mechanism to allocate these interns should consider the will of faculty (autonomy), merit, and availability of funds. These interns will be our most likely future PhD students.

Such outreach programs would be mammoth in size. They require a dedicated office with professional staff serving equal duty to convince and cajole faculty to participate and also to reach out to external institutions for recruitment. Faculty needs to be sensitized on the importance of their participation in such programs, to build excellence in the institution. Any right-minded faculty would understand the importance of recruitment to multiply the success of their work. The institution should put in the right incentive structure to encourage faculty to actively participate.

Personal Benefits

Good advertising can only sell a good product. As we have seen, salaries for research faculty are lower than salaries offered by Google for a fresher software engineer. PhD student stipends are equivalent to IT service company salaries for jobs that require comparatively few technical skills. We need to more than double the PhD stipend and increase faculty salaries by 50–100 percent.

The long-term solution is through institutional autonomy and accountability. University leadership should have freedom to determine salaries. If they do not provide what is needed to attract the best, they will be left behind and lose on reputation and funding. In the short term, our institutions need to find nongovernmental sources of funding to supplement faculty salaries. To recruit top candidates, institutions must offer attractive salary at the time of hiring. Supplemental funding could come from industry, alumni, and foundations in the form of endowed faculty chair positions.[9] Our institutions should strive to add 100–200 chair positions over the next five years. These chairs could be awarded to currently high-performing faculty members and also as a lure to attract promising new talent.[10] Similarly, institutions must raise money to double the PhD stipend for the top 25–50 percent of students.

Finding ways to augment salaries and stipends will not be easy. Salary supplements can be viewed with distrust by the government. The government tries to maintain a fine balance of salaries among various public officials. They see any attempt to get add-ons as a way to go around their regulations. Donors on the other hand are happy to invest in buildings, equipment, travel, and even salary for additional faculty and students. But they see the augmentation of salaries for faculty as a "waste."

This way of thinking needs to change. The government should see itself as subsidizing the faculty salary and being one of several donors toward it. In fact, they should encourage private fund-raising and then match over and above what the institute can raise from nongovernmental donors. Similarly, the private donors should consider that the government contribution toward salary is a subsidy—not necessarily the entire deserved amount. The fund-raising office of the institute should take it upon themselves to justify to donors why the salary of a chair professor should match global standards (in consideration of parity). They need to explain that in sponsoring a faculty chair position at MIT or Harvard, they would have spent a sum of 2 million dollars in the 1990s. A large portion would go toward salary with no government subsidy! On the question of supplemental salaries to sitting faculty, the office must clarify that they are funding a position, not a person: by virtue of their gifts, better and better candidates will fill the position in the future.

Salary augmentation may be the most formidable task for institutions. In the short term, however, there are many other things institutions can do to make themselves more attractive. These include well-maintained housing systems, medical facilities, child day care, and recreational facilities such as gymnasiums. These typically exist already, but are often over capacity and not well run. Outsourcing such functions might improve their quality and efficiency. Universities could establish an HR division responsible for campus life satisfaction—listening to grievances and addressing problems. At the end of the day (literally), we want our faculty to feel needed, respected, and cared for.

Environment

In addition to salary and lifestyle benefits, we cannot underestimate the importance of professional environment. Even by doubling faculty salary, we might not be able to lure a MIT or Stanford faculty here. Here, they would not have autonomy nor great students, great peers, a culture of high performance, or a helpful, nonintrusive bureaucracy. Poor salary and lack of personal amenities might be elimination criteria, but not necessarily a selection criterion.

The professional environment builds over time, by accumulation. It is a chicken-and-egg scenario: good faculty attract good students, good students attract good faculty, etc. Our institutions may not have this environment now, but they need to demonstrate that they are mindful of its importance and must demonstrate to potential hires that they have a vision of how to get there fast. If the university leadership shows commitment to this goal, and has a plan that lists definite steps and can show some progress, it might be enough to entice some risk-taking faculty to take a chance. Anyway, risk-taking researchers often produce the best success.

Another way to go about it might be to get at least a few distinguished faculty members to the institution by any means necessary: persuasion, pleading, enticing, or otherwise! These few are a magnet to their fans: young faculty members who are inspired by their work and who have a strong desire to work with them and learn from them. An MIT alumnus at a private research lab in India told me that he chose to come back because his adviser had come back for family reasons. Several of the adviser's students followed. If we can convince such people to come here and spend some considerable time, then faculty and student recruitment will become much easier.

Finally, we should consider our policies on attracting nonnative Indians. We discussed in the beginning of the section that Asia and Africa can be a fertile ground for recruiting research talent. Unfortunately, most of our government institutions scarcely recruit or fund foreign nationals as PhD students or faculty![11] One of the IIT directors told me that the minister says the Parliament will not allow public money to finance foreign nationals. We should learn from the United States here, the land of immigrants. Even in the current political atmosphere, the United States still actively entices the most meritorious to join their ranks. Investing in great researchers means investing in India's future. In terms of economic and social benefits, and the enhancement of the Indian ecosystem, the nationality of the researcher is of no consequence. Science is a great uniter and an even better equalizer.

Connect and Spur Collaboration

Researchers are social people who thrive on interactions, relying on their peers and their world at large to discover what is new and worthy of pursuit. Researchers need to communicate face to face, often, for progress to happen. Tacit knowledge is essential to good research, and it can only be communicated and exchanged in spirited discussions and collaborative efforts in geographical proximity—not through tightly written technical papers or cursory e-mail briefs.

How can we facilitate these interactions? For one, researchers and their collaborators must have easier access to international travel. Lack of travel funding remains a top concern among faculty members and PhD students. Our researchers should be free to travel as much as their projects demand. The long-term solution for adequate travel funding is to allow faculty to secure the money through project grants from government agencies. In this way, travel money will correspond directly to the professor's ability to raise money for his or her project, which is also some indication of the worth of the project and therefore the value of the travel involved. Furthermore, the terms of the grant should not unduly restrict the use of the money for conference travel specifically, but also allow some flexibility for visiting and inviting collaborators and guest speakers.

How much money should the agency allow for a given project? The onus on justifying the travel required falls on the proposal writer. However, it must be considered without prejudices associated with foreign travel grants. Travel is necessary in the work life of a true researcher. By my back-of-envelope calculations, a high-performing researcher and his or her students require more than double the current level.

In addition to supplementing travel funds we must also lower the bureaucratic regulations and processes surrounding disbursement.

The government and the institution often place additional checks and balance on disbursement, even after approving the travel money. For example, they might necessitate getting an approval from a committee or the institute director for each individual trip. Once the funds are allocated, it is counterproductive to put such additional checks on the process. Researchers should be able to access the money without extra interference. Similarly, restrictions on airlines (such as the directive to fly Air India) must be eliminated. The government should rather create incentives for flying Air India—perhaps in the form of bulk discounts—not mandate it. For visits of foreign researchers, we need to ease the process of getting a visa.

Establishing travel money as a component of project grants may take time. In the meantime, we can find some interim solutions. For example, currently our institutions allocate travel money to each faculty equally. Meritocracy is a better system for allocation, based on factors such as project money raised, number of active PhD students, peer-review ratings, and so on. High-performing researchers deserve a greater portion of available travel funding for the simple reason that they make more productive use of the funds. Similarly, PhD students should also be subject to the merit system. If a student travels one time to an international conference for a great paper, he or she should be assisted with more funding, not eliminated from consideration because he or she used up his or her single allotment.[12]

Governmental and institutional funds need to be made available for inviting international collaborators and guest speakers, together with mechanisms to give it out meritocratically. More importantly, providing reasons why any star researcher would want to travel here is the tough part. What does our present research ecosystem offer to world-class researchers? One possible incentive is money. China and Singapore among others have delivered truckloads of money to institutions such as MIT (and the faculty) for their faculty to come and participate in various programs. I would say we should not buy the vanilla money option, rather combine it with innovative programs. For instance, create a pot that researchers from India and abroad could access jointly for research purposes. The Indo–US Science and Technology Forum operates in this way.

World-class researchers will visit India of their own accord if they think that their investment in time will pay dividends in their research. They can come for our unique research work, problem areas, data sets, forums, and lower costs. A "me-too" approach will not get us far. Consider that the major conferences and workshops were conceived and took place outside India most of the times. Replicating conferences and workshops in similar areas is useless and will garner no interest. But are there gaps to fill? For example, discourse in the area of assessments (my own specialty) mostly takes place in the pages of journals. The field has hardly any conferences. This is an opportunity to take the lead by raising funds and setting up a global program committee to organize and run it. Another example is Learning@Scale conference launched in 2014 buoyed by new research questions and answers MOOCs, such as edX and Coursera, offered. We can find a new or underserved area as well. If we can create recognizable value in our work and forums, the global community will not require extra incentives to come visit us.

Geographically, we need to "look East" in our global collaboration strategy. China, Singapore, Hong Kong, and South Korea have some very highly regarded universities. These countries are in near geographic proximity and have lower cost structures than the European and American universities. They are also more hungry for growth than their Western counterparts are. Enacting programs that could spur people exchange, talks, visits, and collaborations would help us widen our scientific research ecosystem.

Of course, collaboration is not merely an international affair. We should seek to spur collaborations and connections among our own Indian researchers. Simple things as shared cafeterias, interdepartmental luncheons, and various types of social events at our institutions can go a long way. Furthermore, we should not underestimate the importance of informal interaction between faculty and PhD students. Encouraging informal interaction is important in that it encourages and leads to formal collaboration.

Interuniversity centers, conferences, workshops, and summer schools can help spur collaboration among institutions. One novel idea is the National Mathematics Initiative. Established by SERB and led by IISc, it seeks to spur interdisciplinary research through workshops, summer schools, and compact courses. Indian and international researchers alike participate. It chooses a theme each year and then accumulates effort around it. If we can multiply such initiatives as this, under the direction of faculty directors who are enthusiastic, engaged, and well supported financially, then we will witness another positive step in the right direction.

Bring Speed and Efficiency

Researchers work best when they can work with speed and efficiency. One major roadblock is the difficulty of procuring and accessing instruments and materials. Most often, the reason is the bureaucracy with their associated committees and tenders. It all needs to go!

Excess of regulation stifles performance. Researchers should be entrusted with the independence to spend their allotted equipment funds with speed as needs arise, with the understanding that their purchases are subject to audit. Institutions need to establish procurement and financial diligence team in place, such as exist in many global corporations. They should monitor any instances of faking purchases or doctoring receipts and take due action as required.

In the field of research, the wisdom of buying any particular instrument at a particular cost is a tricky arena. The question of value is highly ambiguous in the world of super-specialized instrumentation, given the subjectivity in instrument performance. Furthermore, remember that much research takes place through trial, error, and hacks in instrumentation. Identification of the most suited instrumentation must be left to the researcher himself or herself out of deference for the trust that has been placed in him or her by the grant-making body. Additionally, an institution could establish a peer-review system to determine what is within the norm and provide feedback to researchers.

Such a system with procurement and financial diligence team will work well. Oftentimes, the mere existence of such bodies serves as a sufficient deterrent to fraud. Besides, even absent such systems, the waste or misappropriation of funds will be reflected in the researcher's work, thereby jeopardizing future grant awards. And let us never forget that the most powerful deterrent of all is the loss of respect the researcher would suffer in the eyes of peers.

The system may still experience leakage. As people will never be perfect, we must accept a certain degree of fraud. If leakage remains around 5 percent, it is a small price for the benefit of the 95 percent who are pursuing their work conscientiously and creating value for the economy and society. Anyway, why should we assume the government oversight boards are not themselves fraught with corruption? Are we confident that the multiplicity of bureaucratic committees and overregulation has actually reduced waste, or even slowed it down?

Researchers have difficulty procuring instruments, but they also have difficulty in using the instruments that are already available. This is due to the lack of well-trained staff who operate and maintain the instruments. Well-paid and skilled technicians are indispensable for the efficient use of sophisticated instruments. Without such personnel, institutions cannot even realize the ROI on the purchase of the instrument. They ensure ease of access and ease of use to the entire university community. Without such personnel, use of instruments is often limited to one or two faculty with some preexisting knowledge of the materials. Additionally, a trained staff prolongs the life of the instrument and learns the nuances of the instrument to use it more effectively and for various purposes. They free the researcher from having to divert his or her time away from his or her research in order to learn how to operate the machinery himself or herself.

Last but not the least, the access to instruments should be widely open. In fact, many university instruments sit idle or are used 10 percent of the time at most. The instrument should be made accessible to all faculty and students in the institution. They should be able to block times to use the instrument. It should be also opened for commercial usage and academics at other institutions at a price. In case of excess demand, any of the regular prioritization methods may be used which takes into account precedence, purpose, and volume. This constitutes a potential untapped revenue source.

Create Research-focused Universities and Leaders

We need to create great research institutions. They must be laser focused on becoming research leaders, endowed with all the resources they need to execute efficiently. All the interventions discussed until now can only be implemented through such institutions. We can reinvigorate the institutions we have already and also build new ones. It is not merely wishful thinking: looking around we can see examples of both. In China, SJTU has existed since 1896. Then in 1998, the school underwent a transformation that now places it among the top 150 institutions in the world, with a ranking of 16 in engineering. HKUST was formally launched in 1991 and is today the 31st university in engineering. In science and social science, it ranks between 101 and 150.

If our institutions are to excel to world standards, they must be led by individuals whose key focus is to advance research. This goal should be their prime motivator. They should not be bogged down by day-to-day administration such as coordinating teaching, regulatory reporting, facilities management, and admissions. This is where a lot of time of our university leaders go today. A strong leader must delegate the administrative tasks required to maintain status quo while he or she devotes his or her time 24/7 to pondering and advancing the vision around great research. In business terms, the relationship between the leader and his or her subordinates should be like that of the CEO and the COO. Building new research programs, recruiting top research faculty, building research centers, raising money for research, and institutional collaborations—these things exclusively should be his or her domain.

And he or she needs to "own" them by accepting both responsibility and accountability.

These leaders must be able to articulate clear goals of where they want the school to be in 5, 10, and 20 years. Such goals can be a mix of outcome and input metrics. Outcomes can be citations, disruptive research results, IP generation, industry collaborations, press coverage, awards, and recognitions. Input metrics can be the number of schools, departments, annual budgets, funding raised, profile of faculty, number of international students, number of international trips, international collaborations, and research facilities. Many of these things are subjective evaluations, and not easily quantified by numbers.[13]

The university requires periodic research evaluations and benchmarking to determine how well it is achieving its goals. Such evaluations can be done through a combination of self-evaluation and evaluation by the university management, as well as external reviews, both from academia and industry. Evaluations should follow global standards, though informed by our local needs and focus areas. Lately, India prefers to reject global standards, whenever our institutions rank very low according to them. Programme for International Student Assessment (PISA), a test of school student achievement worldwide, recently gave India a low score, as did an international measurement of world universities. India responded that the tests do not align with India's circumstances. I would agree that benchmarking methods must be constantly reevaluated and adjusted, and must take into account local conditions. However, criticisms and adjustments must be undertaken scientifically, not politically. I doubt India has such a scientific proposal. The global parameters are 80 percent good enough.

Evaluation and benchmarking should be the domain of an exclusive university office that reports directly to the university leader. The university board as well as other stakeholders should have ready access to these reports, so that they cannot be censured by the leadership. The importance of such an office cannot be overstated. By these measurements, we will hold the university leadership accountable for performance. We have a good example in SJTU. It established the Office of Strategic Planning (originally the Office of Policy Studies in 1999) to take on these responsibilities and support the university to become a world-class institution.

The final ingredient for spurring research progress in India is "research in research." We should embark upon a continuous scientific movement to understand how to best align our institutions to achieve scientific advancements useful both to ourselves and the world. This book is one contribution to such a movement. We need to learn from the examples from all the world-class universities across the globe. Also, we need to take account of our own particular needs, challenges, and strengths—through continuous experimentation, pilots, and analysis of results, we will learn how to move forward with our scientific agenda.

Notable examples include Philip Altbach from Boston College who has done substantial work in international higher education, some concerning India. Jonathan Cole of Columbia University and Henry Rosovsky of Harvard University have written books on the new American University and how to run a university, respectively. The Graduate School of Education at SJTU did substantial work on Chinese science and technology policy in the early 2000s, including how to develop world-class universities. Their work led to the Shanghai World Class University Ranking System, which has become a global standard today. India requires a similar concentrated research effort in deciding our science and research policy. Funding agencies, university leadership, and faculty need to work together to make this happen.

And while we think about the big picture, we must begin putting the bricks in place. A successful, future-oriented university is a well-structured university, with professional offices for industrial outreach, marketing, fund-raising, and alumni affairs. These tasks are critical and cannot be the domain of a few professors or devoted alumni. They require dedicated offices staffed by professionals who are answerable to the university leadership for their results. At the top of these efforts sits the university leader. These engines of growth must be his or her charge.

Nongovernmental Action

The Indian institutions that we have been discussing are public institutions. Harvard University, MIT, and Stanford University are all private. One professor remarked to me that if one private university in India could show the way, it would transform the Indian research ecosystem forever. I could not agree with him more.

There have been some serious efforts at private higher education in India recently. However, most have focused on teaching, while others seek to fill subject matter gaps such as in social science. None of them was designed for pioneering global research. An Indian private research university is an idea whose time has come. Three to four experiments must start, and then some of these will succeed. I feel confident that there is sufficient interest from wealthy private individuals within India as well as Indians abroad who would support such a school. We are merely waiting for such higher education entrepreneurs to emerge and take the reins. In this book, he or she will find the justification and the blueprint for making it happen.

Similarly, there is opportunity to create world-class private research laboratories. IBM Research Labs, Microsoft Research Labs, and the erstwhile Bell Labs, have produced great and impactful innovations. They run on huge budgets: Microsoft Research Labs is estimated at $500 million each year and IBM Research Labs at a billion dollars. Some Indian companies have attempted to establish private labs, and some labs have attempted to extend their operations here from abroad, but these efforts have not seen enough success. These need to be reinvigorated to become world class. We need to learn from the recent nonuniversity private research efforts like SpaceX, which have succeeded in doing the impossible.

Further, private players have intervened and impacted in areas such as policing, food security, and primary education, which are traditionally in domain of the state. They are capable of similar impact in progressing research. They can help with evaluations particularly. They can be watchdogs and advocates. India could use an Annual Research Status Report similar to Pratham's ASER and

Aspiring Minds' employability report. Such evaluations would have to be based on data, and fortunately, substantial relevant data is already available in the public domain. Based on its findings, the organization can exercise its advocacy role by exerting pressure on the state and institutions for reform. The organization can also act as a rating agency providing feedback to individual institutions, driving competition, and helping optimal allocation of resources by merit.

Private organizations can provide similar services in consulting for research benchmarking and policy interventions. Such services would be useful for government universities to continuously improve, and private universities that are unsure about how to go about establishing research programs, even if they want to. When I was at Queensland University of Technology (QUT) in Australia, I viewed a presentation from an education-consulting firm that was engaged in this type of work. In fact, QUT had hired the organization to rate it on various indices of excellence in higher education.

The scope of potential activity for private players is limited only by the creativity of individuals. Here are some more examples. Private actors can help research philanthropy. They can help donors find the best programs for their money and interest. Many such organizations operate today in the CSR space, helping funding organizations find the right NGOs. Private actors could create dedicated media companies around Indian science, whether newspapers, magazines, apps, or websites. Other companies could help identify and place PhD students and faculty. The possibilities are endless.

Furthermore, philanthropists can do more than just provide research funding. They can help create incentives and shape the market. Although these functions are also considered the domain of the state, philanthropy can use the power of wealth to influence change. They could create awards and provide monetary incentives for great research. They could help create social respect for researchers by promoting star achievers and recipients of funding through public talks and interviews. They can enable the ecosystem. They can do so much more than just write checks.

Promote Science Entrepreneurship

Indian companies should be second to none. We do not need to merely copy Western models. We can have original ideas that reach the global market. We can create innovative new companies based on true scientific and technological advancements. I urge my entrepreneurial brethren and entrepreneurs-to-be to take up science entrepreneurship. With sound business sense, they will discover a large market, greater value, greater global appeal, and higher returns, than if they continue to pursue the same old, same old.

We need to connect PhD students to entrepreneurship. Currently we steer primarily undergraduates to entrepreneurial careers. A better focus would be on those who actually do research. We could have business and entrepreneurship courses within the PhD curriculum. Departments could host talks with investors and entrepreneurs at which PhD students could showcase their work. We should also encourage PhD students to undertake internships as part of their PhD study.

These efforts will build a bottom-up supply of interested individuals. But the ecosystem needs to respond. This is by providing funding for science entrepreneurship by investors who value innovation, respect the research process, and are patient in undertaking long-term, risky propositions. High net worth individuals should establish funds, incubators, and accelerators for innovative companies. If we can achieve one great success story, it will inspire the market to follow. Unfortunately, today the world seems to believe that we do not create next-generation start-ups. If we have the courage to invest and try, we can change this perception.

Creating competitions and awards for innovative companies can help. India does have forums for "innovative," but these mostly cater to innovation in the business sense. We need competitions that recognize scientific and technological innovation. Competitions excite people to tread unknown territory—and if the competition holds out awards of funding, there will be great response. They also bestow important social recognition and succor to entrepreneurs who are dedicated for the long haul.

Science start-ups are an essential part of the research ecosystem. They are integral to realize the economic and social benefit of research. They show the public and the government the value of research, which helps build the case for continued support.

Find New Creative Ways

So far in our discussion of how to improve the research ecosystem, we have focused on proven measures of success for which there exists strong evidence. Other countries have faced problems similar to us and have overcome them to create great research programs. While all these solutions hold promise, we are not limited by them. Numerous other possibilities exist for innovation in the pursuit of strong research programs. We can give the world new ways of doing research more successfully and develop unique methods and strategic initiatives based on our unique strengths.

In this section, I will let my pen loose and toss off several such ideas and suggestions. Some of these ideas might sound half-baked, but they might just ignite a spark.

The New PhD

In the last two decades, access and communication have become infinitely faster thanks to new communication technologies. We no longer have to wait weeks or months for the latest research literature, spend extensive time manually marking corrections in documents, or for correspondence from colleagues. Regardless, the duration of the PhD degree has not decreased, nor has the quality of the PhD thesis improved. In fact, in the field of computer science, I find that theses written in the 1980s are deeper than many of those written today. It seems we are set in a system that is resistant to change. For some reason, we begin with the assumption that the PhD will take 4–6 years, and then the system works to fill up the time.

But is it all necessary? More and more people think not. Can Indian institutions demonstrate that PhD can take less time—say three years—without compromising quality of work? If one institution were bold enough to take on this challenge, they might just prove it to the world. For ourselves, it will be greatly helpful since the long duration of the PhD degree is a top deterrent for Indian undergraduates. This experiment is an opportunity in waiting.

Here is another. Today, substantial research happens in industry. In my company, I employed at least two people whose research output and results would have earned them a PhD from a world-class institution if they had performed the work there rather than here. But of course, no university will award them academic credit for these efforts. They cannot do research and teach in a university without a PhD.

Why not have a "fast-track" PhD program for people who have performed substantial research—enough to fulfil the typical requirements—outside of the university environment? In a year or two, they could satisfy their coursework and write their thesis. It is a win-win. The individual can move ahead in his or her career, industry can attract smart researchers, and universities can get a stronger supply of PhD students. These are just two ideas to create a smarter and more outcome-oriented PhD program.

The New Citations

We traditionally measure research impact by citations. However today citations come in many forms: dissemination aka Twitter, LinkedIn, one's homepage, and networks specific to research such as ResearchGate. These have become favorites of industry. We can aspire to develop new ways of measuring impact that take into account number of tweets, number of downloads, number of likes, etc. This type of communication may have a large impact on people's lives. Here is an opportunity to take lead in creating such impact and measuring it.

Let us also consider the research paper itself. Are there new kinds of papers, just as there are new forms of citations? Already we see how the presentation of content has changed in newspapers, TV, and radio. Should research papers change too? In academia, many conferences in data science and computational biology request data sets and algorithms for verification. A leading data science conference, KDD, requires a short video on paper. Authors compete based on the number of views on their video on YouTube. These are incremental steps but suggest that the time has come for a change. What other new ways can we come up with to present research results for better and wider consumption? Could we use ideas from design thinking?

Network of Private Research Colleges

India has 3000+ private engineering colleges. Today, these focus primarily on teaching. Can they undertake research as well and become viable options for our high-quality PhDs, within their current budgets? They can if each chooses one research area in which to specialize and attracts 5–10 high-quality research faculty. They can choose their areas according to their strengths: geography, tradition, current faculty specialty, or connections to industry. This is a small measure easily achievable within their budgets (costing perhaps a couple of crores annually—roughly $300,000) and could create real output by realizing a critical mass in a single field. Together, they could accommodate 30,000 PhDs in research. Such a program would have a positive effect on teaching quality and reputation, and would inspire more undergraduates to pursue PhD study. Furthermore, these colleges would comprise a network, collaborating with each other on multidisciplinary research. The many colleges I talk to are looking for new things to do. Some would embrace this idea enthusiastically. But they need competent guidance.

The New Research Places

I have often wondered why we need large sprawling campuses to house our new research places. Private labs exist in compact buildings in industrial areas. MIT exists like a set of buildings within the city. PhD students are typically blind to the campus and only see the path from their dorm to their lab! We should consider housing new research institutions in more compact spaces, which offer both functionality and savings. This can help us scale fast and invest in only what we need to get the maximum bang for the buck. Can government change their regulations regarding the physical campus? Can we experiment with private setups that do not require government approval? We can more easily change government regulation if we can first offer an example of efficiency and success. And here is another thought: what is the structure of the new research place in our age of mobile communication and virtual reality? I suspect that the school of the future would be unrecognizable to us today.

Crowdsourcing Innovation

We find it pays great dividends to ask the crowd—the people at large—how to solve a problem. People engaged in other fields can sometimes solve a problem that confounds practitioners in the particular domain. This should not be surprising, given that they bring a fresh perspective and experience from their own field to the problem. The wider community is great in generating different ideas and approaches to problems that the experts can then develop into full solutions. This is not the traditional way of doing things in the typical research institution. Instead, a single researcher along with his or her student thinks about a problem and attempts to solve it. They do not involve outsiders. At Aspiring Minds, we have begun involving the "crowd." For every new problem we pick to solve, we do a hackathon/competition of kinds. Participants from multiple teams generate good ideas and then the research engineer involved takes one or more to fruition. Can our institutions find mechanisms to involve crowd intelligence in the work of our researchers, perhaps by providing digital platforms and services? Can they determine issues of attribution and IP in such cases? By reaching beyond traditional knowledge, we could probably do better and faster research, with more chances of breakthrough.

Equipment Manufacturing and 3-D Printing

India has connected strengths and weaknesses. If we can learn to address our weaknesses by applying our strengths, we will find ourselves on a fortuitous path. For example, we can build things at low cost, but our researchers suffer from lack of components and materials. Today, 3-D printing is revolutionizing how we think about lab equipment. You no longer buy the equipment, you print it! Today, you can print car and airplane parts, surgical instruments, and organs! Can India take a strategic bet in 3-D printing technology? If successful, we could democratize equipment availability for the world. If we developed a core competency in this technology, we could help researchers everywhere. This could provide a great fillip to applied research by making availability of customized components fast and with quality.

I am confident that India can become a leader in science and technology. This is the most opportune time to reinvigorate our research agenda—it is now or never. We need to take ownership, set our goals, and intervene according to the stated design principles. This is our responsibility to India and to humanity.

Notes and References

Preface

1. See http://www.worldbank.org/en/news/infographic/2016/05/27/india-s-poverty-profile (accessed on September 18, 2017).
2. See http://ghi.ifpri.org/ (accessed on September 18, 2017).
3. See http://hdr.undp.org/sites/default/files/2016_human_development_report.pdf (accessed on September 18, 2017).
4. See http://www.firstpost.com/india/indian-science-congress-is-a-circus-wont-attend-it-nobel-laureate-v-ramakrishnan-2572268.html (accessed on September 18, 2017).
5. *Jugaad* (or "hacks" as they are known in the West) are quick fixes or work-arounds meant for quick functionality, when building new things.
6. This is based on total number of citations from 2012–2016.
7. I use "university" to include universities, colleges, and degree-providing institutions throughout the book. Clarification about their difference is provided in Chapter 9.
8. See https://www.technologyreview.com/lists/companies/2017/ (accessed on September 18, 2017).
9. In 2017, as this book goes into production, MIT TR has included Flipkart from India in their list. MIT TR may have changed their evaluation, but I have not! I see no evidence of any scientific or technological innovation made by Flipkart.

Acknowledgments

1. Please refer to sciencesatyagraha.com for a more complete list of people I talked to.

Chapter 1

1. Several companies today claim to be building a driverless car. Some of these include Nvidia, Tesla, Apple, Uber, and General Motors.
2. One could just imagine how high this rate would be in India—probably 1 in 2 cancer patients!

3. Innovation refers to innovation in science and technology, until otherwise stated.

4. I use "university" to include universities, colleges, and degree-providing institutions throughout the book. Clarification about their difference is provided in Chapter 9.

5. Mariana Mazzucato, *The Entrepreneurial State* (UK and USA: Anthem Press, 2011).

6. Other than universities, research also happens at research labs.

7. However, more often now than before, research results are protected by patents so that a person must obtain permission from the holder before they can even read the study. The patent holder could be the professor, the students, university, or the funding entity.

8. Research results are not always public good—they are protected by patents, more often in fields like biotechnology.

9. A good survey of this literature can be found in the 2004 chapter of Audretsch and Feldman. See David Audretsch and Maryann P. Feldman, "Knowledge Spillovers and the Geography of Innovation," in *Handbook of Regional and Urban Economics*, 4 (2004): 2713–2739.

10. A 1997 BankBoston report (see http://news.mit.edu/1997/jobs, accessed on September 19, 2017).

Chapter 2

1. Even a price war will not yield much advantage, due to the low marginal cost of replicating either product. Besides, the superior product can fight any pricing war more effectively.

2. Another set of companies that are hard to replace are those with "network effects." Research and innovation, which create new efficient ways to form networks and derive value from them, may disrupt them.

3. Some of our companies are able to charge a premium for their services but the premium is marginal, not exponential.

4. Some of our more mature companies are attempting to adopt new methods. But they will likely be followers, not leaders.

5. MIT Technology Review, "China is Building a Robot Army of Model Workers," April 26, 2016, https://www.technologyreview.com/s/601215/china-is-building-a-robot-army-of-model-workers/ (accessed on September 19, 2017).

6. Consider the Literature survey in Sakiru Adebola Solarin and Yuen Yee Yen, "A Global Analysis of the Impact of Research Output on Economic Growth," *Scientometrics* 108, no. 2 (2016): 855–874.

7. Ibid.

8. Ronald Kumar, Peter Josef Stauvermann, and Arvind Patel, "Exploring the Link Between Research and Economic Growth: An Empirical Study of China and USA," *Quality & Quantity* 50, no. 3 (2016): 1073–1091.

9. Jang C. Jin and Lawrence Jin, "Research Publications and Economic Growth: Evidence from Cross-country Regressions," *Applied Economics* 45, no. 8 (2013).

10. R. Inglesi-Lotz, T. Chang, and R. Gupta, "Causality Between Research Output and Economic Growth in BRICS," *Quality & Quantity*, 49, no.1 (2015): 167–176.
11. Lee et al. had also found a causal relationship from research output to economic growth for India. See L. C. Lee, P. H. Lin, Y. W. Chuang, and Y. Y. Lee, "Research Output and Economic Productivity: A Granger Casuality Test," *Scientometrics*, 89, no. 2 (2011): 465–478.
12. R. A. Mashelkar, "India's R&D: Reaching for the Top," *Science* 307, no. 5714 (2005): 1415–1417.
13. This might be true if the marginal cost of doing higher impact research foliowed an increasing function. For example, you need to build a hugely expensive cyclotron to do superhigh impact physics research, whereas you could perform moderately impactful physics research with simpler and less expensive instrumentation. Also, higher impact projects could be riskier. They fail more often than not, resulting in wasted money. This further increases the average cost of research.
14. See http://timesofindia.indiatimes.com/india/36-of-scientists-at-NASA-are-Indians-Govt-survey/articleshow/2853178.cms (accessed on September 19, 2017).
15. See http://www.iie.org/Services/Project-Atlas/United-States/International-Students-In-US#.V-uox4h97IU (accessed on September 19, 2017).
16. People use various spellings for J. C. Bose. I have used the most popular one today. He used Jagadis Chunder Bose in his papers and patents.
17. These numbers are discussed in detail in Chapter 3.
18. Science does modify our understanding of morality and ethics, and may change our goals. This is beyond the scope for our current discussion. But for the curious cats, I will let this thought ignite in your mind.
19. We must also understand the limitations of science. The greatest scientists do so. As a society, we should continuously enlarge the scope of science. At the same time, however, when science does not help, we should be courageous enough to follow our beliefs and our gut; shoulder tons of positive work; and keep trying to turn the wheels of progress.
20. The purpose of this book is to help see the direct connection between scientific progress and university research.

Chapter 3

1. Read a detailed report here: http://noragging.com/analysis/CR2007_05-16_RaggingInIndiaSummary.pdf (accessed on September 20, 2017).
2. *The Telegraph*, "State of Talent," 2012, https://www.telegraphindia.com/1120117/jsp/jobs/story_15013936.jsp (accessed on September 20, 2017).
3. Now called the NASA/ESA Conference on Adaptive Hardware and Systems.
4. J. C. Bose, "On A Self-recovering Coherer and the Study of the Cohering Action of Different Metals," *Proceedings of the Royal Society* 65, no. 416 (1899, April): 166–172.

5. We use data from Scopus and SJR (based on Scopus data) as the basis for most analysis in this chapter. As stated in different sections, public data about conferences, journals, and Microsoft Academic rating are other data sources used.

6. In 2014, there were reportedly 12 journals from Beall's list in the Scopus' database.

7. See http://databank.worldbank.org/data/reports.aspx?source=Education-Statistics:-Education-Attainment&preview=off (accessed on September 20, 2017).

8. Aspiring Minds' National Employability Report shows that 80 percent of engineers are not employable for any technical job, while more than 90 percent of science and commerce graduates do not have conceptual understanding of their subject matter.

9. Thomas Barlow. *Between the Eagle and the Dragon* (New South Wales: Barlow Advisory, 2013).

10. In some sense, we become the best judge of our works! It needs to satisfy us and our intellectual self.

11. Varun Aggarwal, "Where Does India Stand as Machines Become Intelligent," LinkedIn, November 18, 2015, https://www.linkedin.com/pulse/where-does-india-stand-machines-become-intelligent-varun-aggarwal (accessed on September 20, 2017).

12. For detailed counts per journal/conference, refer to "Data" tab on sciencesatyagraha.com

13. For detailed comparison, refer to "Data" tab on sciencesatyagraha.com

14. We will come back to a more methodical comparison of different fields later.

15. See http://www.purplemath.com/modules/linprog3.htm (accessed on September 20, 2017).

16. These are mostly reproduced from Narlikar's work with some small changes.

17. J. Bose, "On the change of conductivity of metallic particles under cyclic electromotive variation," (originally presented to the British Association at Glasgow, 1901, September).

18. See http://www.thehindu.com/news/national/has-iisc-contributed-to-society-narayana-murthy/article7426651.ece (accessed on September 21, 2017).

19. Published in C. N. R. Rao, *Current Science* 109, no. 5 (2015, September 10): 844.

20. See http://www.thehindu.com/todays-paper/tp-opinion/revamping-indias-scientific-ecosystem/article7493306.ece (accessed on September 21, 2017).

21. Professor Deb has 100,000+ citations.

22. Abhay Bang, "Meeting the Mahatma," http://nirman.mkcl.org/Downloads/Articles/Meeting_the_Mahatma.pdf (accessed on September 21, 2017).

23. Varun Aggarwal, "My Gandhi" (unpublished).

24. This does not purely measure the absolute growth of citations in a single country. A growing country could seem like it is actually regressing if an extremely fast-growing country in present, affecting the world average. It is more like a rank, where a country may improve but still decay in rank because

others improved more over the same time period. For all practical purposes, this does not matter for our inferences.

25. The contribution from the United States has gone down from 32 percent to 23 percent, largely due to China, which aspires to compete with the United States.

26. Refer to "Data" on sciencesatyagraha.com for detailed analysis.

27. Systematic error is an error which affects all entities compared similarly and thus becomes a "don't care." For example, the zero error in a weighing machine is a systematic error. When we compare the weight of two people, it is added as a constant and doesn't matter.

28. Some of these may have Indian contributors given they are open source. But none was initiated or led by people in India.

29. This list is in no way exhaustive, but a reasonable reference point. Our data analysis is based on the snapshot of the page on November 16.

Chapter 4

1. The PhD degree is discussed in detail in the next section.

2. Department of Science and Technology, "Full Time Equivalent of Manpower Employed in R&D Establishments as on 01.04.2010", 2010, https://data.gov.in/catalog/full-time-equivalent-manpower-employed-research-and-development-establishments (accessed on November 16, 2017). This data is not available beyond the year 2010.

3. These numbers include people with and without PhD degrees. According to the 2005 government data, in government labs, 26.5 percent of these researchers were PhDs and 41.4 percent had a postgraduate degree. Such data is not available for the industrial sector or for higher education institutions, or for the years beyond 2005. Data for government personnel is expected to become available in the next survey.

4. This analysis would be sharper if we had counts for the total numbers of researchers with PhDs in each country; however, that data is not available.

5. A country can amass a high citation count just by having a large number of researchers, even if those researchers are individually low performing. This is not satisfactory. If 2,500 researchers garner five citations each, they do not equal the impact of the discovery of helical structures by Watson and Crick (with their 12,500 citations). See J. D. Watson and F. H. C. Crick, "Molecular Structure of Nucleic Acids: A Structure for Deoxyribose Nucleic Acid," *Nature* 171, no. 4356 (1953): 737–738.

6. Microsoft Academic Research is a compendium of 80 million papers that utilizes software-based application programming interface (API) access for downloads. They provide data amenable to our productivity analysis.

7. We rearranged the data by authors to find their number of papers and citations. Using automated and manual algorithms, we matched the authors' affiliated institutions to India, China, the United States, the United Kingdom, and others.

8. We choose exponentially increasing citation counts, since citations to authors follow a power law.

9. In the last section, the analysis indicated that China has a lower proportion of good quality researchers, whereas this section clearly shows that it is higher. I attribute this to inflation of researcher numbers for China. The 1.6 million researchers reported are not a right indicator for paper-producing researchers.

10. We include Australia among the 7 countries considered earlier. Data for other countries were not easily available.

11. We will see later that only a quarter of the top students in colleges beyond the top 50 are interested in pursuing research in India.

12. We used language skills, aptitude, and domain skills as criteria for identifying top undergraduate students. We sent the survey to the top 15 percent of engineering students based on their Aspiring Minds Computer Adaptive Test (AMCAT) scores in their final year of study. AMCAT is India's largest employability test. It is taken by 2 million students every year and used by 3,000+ companies for recruitment every year.

13. For further details of the sample, refer to "Data" tab on sciencesatyagraha. com

14. The high number for students opting for a PhD abroad could be because the response might be a proxy for the student's desire to go to the United States. It could also be due to self-selection as our survey was titled "Research as a Career." This wouldn't disturb the ratio of students who opt to do PhD in India versus the United States. It may just mean that the people not interested in research could be higher than 35 percent.

15. Refer to "Data" tab on sciencesatyagraha.com for detailed results.

16. We will discuss more about career options a little later.

17. Refer to "Data" tab on sciencesatyagraha.com for detailed table. People wrote open-ended answers to this question. We then categorized their responses to arrive at these statistics. A single answer could encompass more than one reason.

18. People could choose more than one option.

19. Refer to "Data" tab on sciencesatyagraha.com for detailed table.

20. GATE has gained a little more popularity today after public sector undertakings (PSUs) started using it for recruitment. Government jobs remain a draw in India.

21. GATE is for engineering disciplines and NET is for science and humanities.

22. To verify the claim of students who reported to have read one or more papers, we checked whether the titles they named were roughly correct. In Table 4.5, we report the total percentage of people with valid responses. We consider those who answered "No" and who did not provide titles as "Did not read." Detailed table is available under "Data" tab on sciencesatyagraha.com

23. For detailed table, refer to "Data" tab on sciencesatyagraha.com

24. IT industry pays more than most other science/engineering degrees.

25. Many PhD students in the United States aspire to a tenured faculty position. MIT consistently warns them that given the number of research faculty

positions available annually in the United States, only a slim minority of them can expect such a placement. Can this be an opportunity for India to hire some top talent?

26. For detailed table on the questions we posed to Indian PhD students abroad and their responses, refer to "Data" tab on sciencesatyagraha.com

27. For completeness, this is what we learned from surveying Indian PhD students. For PhD students in top Indian institutions want to opt for a career. Only a third of them want to stay in India. Around 58.4 percent want to go abroad for research and then come back, while 5.6 percent want to look for permanent positions outside India.

28. See http://www.iitrpr.ac.in/advertisement-no-iitrprfacrectoff-shore2016 (accessed on September 25, 2016).

29. I can personally testify for his research quality. But more importantly, it has been published in top journals in neuroscience.

Chapter 5

1. R. A. Mashelkar, "What Will It Take for Indian Science, Technology and Innovation to Make Global Impact?" *Current Science* 109, no. 6 (2015): 1021–1024.

2. We will discuss the process of funding research in Chapter 9.

3. I face similar problems with books that are not available in India. This includes *The Great American University*, which I referred for this book!

4. Much of this attitude has been encouraged by the institutions in its wisdom to try to curtail corruption.

5. There are some funds from the Indian government for inviting international speakers, but not much.

6. When necessary, a department may hire additional staff using project grant money.

7. Quality is not the only concern. Indian English has evolved into a unique style—the right and the wrong—that does not conform to Western standards.

Chapter 6

1. Each professor is assigned a minimum of 1.5 PhD students per year, on average. Institutions may vary.

2. Jessica Lin, "Unraveling Tenure at MIT," June 11, 2010, http://tech.mit.edu/V130/N28/tenure.html (accessed on September 26, 2017).

3. University/institute/distinguished professor is the highest title awarded to a faculty member for his or her pathbreaking research achievements. They have exceptional flexibility and freedom in their work and minimal reporting overhead. Currently, Harvard University has 26 university professors, and MIT has 15 institute professors.

4. Filling positions is a constant challenge at top Indian universities. See http://indianexpress.com/article/india/india-news-india/42-in-kharagpur roorkee-39-in-bombay-faculty-short-across-iits/ and http://www.dnaindia. com/india/report-2k-vacant-faculty-positions-in-iits-3k-in-nits-mos-hrd-2279963 (accessed on September 26, 2017).

5. R. Ponds, F. Van Oort, and K. Frenken, "The Geographical and Institutional Proximity of Research Collaboration," *Regional Science* 86, no. 3 (2007): 423–443.

6. Jonathan Cole, "The Great American University, Its Rise to Preeminence, Its Indispensable National Role, Why It Must Be Protected" (USA: Public Affairs, 2012).

7. Kenneth D. Campbell, "Study Reveals Major Impact of Companies Started by MIT Alums," *MIT News*, March 5, 1997.

8. Maryann Feldman, a professor of public policy at the University of North Carolina, has studied the impact of geography on innovation activity and "knowledge spillovers." Several empirical studies suggest that location and proximity clearly influence knowledge spillovers. Other empirical studies use different measures of innovation inputs and outputs, but they arrive at similar conclusions. Another set of studies model and explain the reasons for geographic clustering and the mechanisms of local knowledge spillover. For a survey of this literature, see David Audrestch and Maryann P. Feldman, "Knowledge Spillovers and the Geography of Innovation," in *Handbook of Regional and Urban Economics*, 4 (2004): 2713–2739.

9. Recent article, "China May Match or Beat America in AI" in *The Economist* (July 15, 2017) clearly states US and American leadership in AI (see https://www.economist.com/news/business/21725018-its-deep-pool-data-may-let-it-lead-artificial-intelligence-china-may-match-or-beat-america [accessed on November 16, 2017]).

10. Geographical proximity cannot be a substitute for institutional relations. I was told that a government lab is situated directly outside Birla Institute of Technology and Science (BITS), Pilani, but there is no collaboration between them. Strong institutional and informational proximity can help mitigate geographical proximity. Workshops, conferences, and summer schools are some examples of mechanisms that help compensate for distance. Professor Ravindran and Professor Jhunjhunwala from IIT Madras and Professor Sundaresan from IISc have begun such initiatives.

11. One possible explanation is that the less ambitious self-select themselves to join our institutions. This is discussed in detail in Chapter 3 (Research in India: The Past, Present, and Future).

12. Some folks remark that this is part of the hierarchical society in India. I disagree.

13. See http://me.sjtu.edu.cn/english/Xueyuan/Default.aspx?cid=10 (accessed on September 26, 2017).

14. "Excellence in Scientific Collaboration," a section in the *OECD Science, Technology and Industry Scoreboard*, 2015, http://www.oecd.org/sti/inno/scientometrics.htm (accessed on September 26, 2017).

15. The survey includes 73 faculty members across IISc, IITs, IISERs, NBRC, and National Institute of Plant Genome Research.

Chapter 7

1. Similarly, researchers who work in quantum computing or new material research are also far ahead of industry.
2. World Bank, "High-technology Exports (% of Manufactured Exports)," http://data.worldbank.org/indicator/TX.VAL.TECH.MF.ZS (accessed on September 27, 2017).
3. Based on UNESCO data, see https://scienceogram.org/blog/2013/05/science-technology-business-government-g20/ (accessed on September 27, 2017).
4. Excluding Lincoln Labs.
5. We tried to get data from IIT Mumbai and IISc also. It was not published in their annual reports, and they did not respond to repeated calls. We believe their numbers are very small.
6. See http://www.universityworldnews.com/article.php?story=201102042227 22977 (accessed on September 27, 2017).
7. Nandan Nilekani has donated to IIT Kanpur, Kris Gopalakrishnan to IISc, and the Infosys Foundation has donated to IIIT Delhi. The Indian School of Business was also founded by businessmen of Indian origin. They actively participate in research.
8. In 2017, as this book goes into production, MIT TR has included Flipkart from India in their list. MIT TR may have changed their evaluation, I have not! I see no evidence of any scientific or technological innovation made by Flipkart.
9. See http://www.business-standard.com/article/companies/it-s-amazon-vs-flipkart-a-look-at-the-e-commerce-dynamism-in-india-117042700514_1.html (accessed on September 27, 2017).
10. Andrew recently moved out of Baidu, as this book went into production. Baidu announced Haifeng Wang (~3,000 citations) as his replacement.
11. The Indian IT industry caters mostly to customers outside India. They have developed their business by building sales and business development teams outside India. This is typically only feasible for companies with a large market size, because of the larger investment requirements and the longer time that it takes to make sales.
12. India also presents a larger market for B2C products. This said, buyers of sophisticated or expensive services are limited.

Chapter 8

1. Prasanta Chandra Mahalanobis is a celebrated Indian scientist and the founder of Indian Statistical Institute. He was at University of Cambridge at the same time as Ramanujan.

2. India poses many unique social science questions such as these. Western researchers are generally keen to work with Indian data sets, researchers, and organizations to tackle them.

3. Claude E. Shannon, "A Mathematical Theory of Communication," *Bell System Technical Journal* 27, no. 3 (July 1948): 379–423 (cited more than 61,000 times).

4. I posted my article "JC Bose: The Real Inventor of Marconi's Wireless Receiver" on the web. The article was later published by *AWA: Ancient Wireless Association Journal* and won the "Best Article" award.

5. Researchers today agree that Bose invented Marconi's receiving device. The question is what constitutes inventing radio. This is subjective and open to debate. I ascribe the credit to both Marconi and Bose, because both contributed scientifically to a critical piece of the transatlantic experiment.

6. See http://www.rediff.com/news/special/did-india-discover-pythogoras-theorem-a-top-mathematician-answers/20150109.htm (accessed on September 28, 2017).

7. Shailendra Mehta is the chairman of Board of Management of AURO University and the President and Director of MICA, Ahmedabad.

8. A larger discussion and treatment of this topic is in Chapter 7.

9. "On Automated Assessments: The State of the Art and Goals," Contributions from Varun Aggarwal, Steven Stemler, Lav Varshney, and Divyanshu Vats, co-organizers, ASSESS, KDD, 2014.

10. Incidentally, MIT's motto is "Mens et Manus," the Latin for "mind and hand."

11. Disclosure: I am biased toward intuitive understanding. I am not too good at the math!

12. Professor Tarun Khanna chaired the commission on entrepreneurship set up by Government of India which informed structure of Atal Innovation Mission (AIM).

13. Find more details on www.datasciencekids.org (accessed on September 28, 2017).

Chapter 9

1. We covered in Chapter 4, how institutions do not do very well in motivating students for a research career.

2. Refer for a detailed history of education, science, and technology policy in India: N. Jayaram, "Beyond Retailing Knowledge: Prospects of Research oriented universities in India" in eds Philip G. Altbach and Jorge Balan, *World Class Worldwide: Transforming Research Universities in Asia and Latin America* (Baltimore: The Johns Hopkins University Press, 2007).

3. Through their research, German universities were instrumental in developing the chemical industry in Germany.

4. Later came the inclusion of professional courses such as business to complete the contemporary potpourri.

5. These comparisons are based on PPP. This may not be ideal, since the equipment for research is available on the market conversation rate mostly. In fact, they are sometimes more expensive for Indian institutions, given transportation costs and custom duties. The ratios given here are on the optimistic side.

6. Other than in the United States and Japan, most world-class universities are public institutions.

7. I take IITs as an example here. Other public institutions have similar autonomy status.

8. Public institutions include some government officials on their governing boards.

9. The IIITs are public–private partnerships and enjoy more autonomy. However, they are responsible for funding their own operating expenses.

10. At one point, the ministry decreed that alumni could not contribute directly to their former schools but had to donate instead to a collective pool. This rule was later revoked.

11. The Indian government also influences private universities and institutions in what programs they can run, salaries they can award, student they can admit, and fees they can charge. Some of these decisions are suboptimal, and the regulatory bodies have been accused of corruption. Private institutions are also in need of greater autonomy as well as accountability for quality. We will discuss this in Chapter 10.

12. See http://www.sandiegouniontribune.com/news/education/sdut-uc-regents-confirm-new-ucsd-leader-set-pay-2012may16-story.html (accessed on September 29, 2017).

13. This is primarily undergraduate tuition. Graduate students are usually funded. Total tuition dollars collected from graduate students does not amount to much.

14. Henry Rosovsky, *The University: An Owner's Manual* (USA: W. W. Norton & Company, 1990).

15. IIT Mumbai corpus was 69 crores, IIT Kharagpur 12.15 crores in 2003. Much of this was buoyed by the fame of IIT's undergraduate education globally. We could not find updated figures in the institution's annual report or by communicating with the institute. See http://www.rediff.com/money/2003/may/24spec.htm (accessed on September 29, 2017).

16. See http://www.business-standard.com/article/technology/iits-remain-big-daddy-of-fund-raisers-bombay-leads-the-pack-111111400007_1.html (accessed on September 29, 2017).

17. J. C. Bose, "Detector for Electrical Disturbances," US Patent No. 755,840, application filed, September 30, 1901, patented March 29, 1904. See http://web.mit.edu/varun_ag/www/jcbosepatent.pdf (accessed on September 29, 2017).

18. J. C. Bose to Rabindranath Tagore, 17 May 1901, in *Archives of Rabindra Bhavan*, Personal Letter, (Santiniketan, West Bengal: Visva Bharati University).

19. Henry Rosovsky, *The University: An Owner's Manual*.

20. CRISPR used loosely to refer to CRISPR-Cas9, a revolutionary gene-editing technique.
21. We have already discussed poor marketing by Indian institutions to attract new faculty and students in Chapter 4.
22. Based on UNESCO data. See https://scienceogram.org/blog/2013/05/science-technology-business-government-g20/ (accessed on September 29, 2017).
23. Based on RTI response.
24. DRDO doesn't come under the RTI bill.
25. See https://research.usc.edu/files/2011/05/Guide-to-FY2014-NASA-Research-Funding1.pdf (accessed on September 29, 2017).
26. See https://www.darpa.mil/work-with-us/heilmeier-catechism (accessed on September 29, 2017).
27. This was confirmed with faculty members.
28. The researcher needs to justify the requirement of the funds.

Chapter 10

1. India needs a strong system, such as the tenure track in the United States, to disincentivize underperformance among faculty.
2. At the same time, we need to reject unscientific and nonevidence-based approaches.
3. For a list of these measures and schemes, refer to "Measures" tab on www.sciencesatyagraha.com. They cannot be considered complete.
4. The Kakodkar Committee has suggested appointing a chief financial officer (CFO) to help manage finances. The Narayana Murthy Committee Report on higher education has proposed asking corporate executives to train university leaders.
5. By this, I admit that all of my own arguments are just one perspective and opinion. I am merely one among many stakeholders involved in our mutual deliberation.
6. Refer to "Leading A Research University" in Chapter 9.
7. In 2017, 42 percent of graduate students at MIT are international students.
8. I was fortunate to meet three remarkable undergraduates through this program, all of whom were intelligent, hardworking, and sincere, and all of whom grew more confident of a research career through their time in the program.
9. An endowment of ₹1 crore could create a 100 percent increment in salary for a faculty member.
10. While industry will naturally have some demands in return for their investment, we must be careful to never sacrifice academic independence for the sake of funds.
11. IITs can have foreign national only as contractual faculty and not in permanent positions.

12. Among high-performing students, priority may be given to those who have traveled less.
13. For example, whether a researcher is a leader in his or her field can only be determined by the opinions of his or her global peers, supported through letters of recommendation.

Index

Index 275

tenure process, 133
US universities, 135
artificial intelligence, 142
critical mass, 143
culture of excellence, 143
ecosystem effects, 141
global community, interactions with
collaborate and recruit, 151
hear and learn, 148
tell and sell, 150
incentives, 129
international interactions, 152
peers, interactions with, 138
promotion ladder, 130
proper training and ongoing skills
enhancement, 130
research direction, 143
research ecosystem quality
parameters, average rating of, 156
research faculty work with PhD
students, 145
researcher
awareness, 89
counting, 73
personal rewards and career paths,
97
PhD, 78
choosing, 85
productivity, 75
professional rewards, 95
quanlity, 83
research faculty, 100
awareness and recruitment, 104
motivation, 101
personal rewards, 105
professional rewards career path,
108
sowing seeds, research, 91
research funding ecosystem, 216
research questions
big-picture thinking, 172
broad and narrow, 174
complete solutions to relevant
problems, 182

consciousness, 173
continued-fraction solution, 172
derivative problem statements, 174
economic and social value, 174
education culture
amplifiers, 188
Atal Tinkering Labs in schools
and institutions, 189
book knowledge, 186
friend predictor, 189
hands-on approach vs. theoretical
analysis, 186
intuition vs. math, 186, 189
intuitive understanding, benefits
of, 188
jugaad, 186
real-world problems, 187
school-level math education, 186
hybrid languages, 174
motivation, 176
open-ended questions, 172
scientific exploration, 173
scientific identity and pride, 178
smaller and narrower, 172

Samsung, 21
science. *See also* science and technology
arrogance of, 48
impact of, 3
innovation, 4
science and technology
accountability, 222
accumulate effort, 234
areas of excellence identification,
232
autonomy and accountability at
private universities, 230
availability of money, 225
awareness creation, 224
Band-Aid solutions, 226
centrality of people, 222
competition creation, 224
connect and spur collaboration, 243
crowdsourcing innovation, 257

About the Author

Varun Aggarwal is a researcher and an entrepreneur. He earned his Bachelor of Engineering degree at the University of Delhi and later dropped out of a PhD program at the Massachusetts Institute of Technology, after having obtained his master's degree. He cofounded Aspiring Minds, an employability assessment company, in 2008 to drive meritocracy in labor markets. He heads research at Aspiring Minds. His work has led to the world's first machine-learning-based assessment of coding skills and the world's first automated motor skills assessment. He has published more than 30 research papers and takes pride in the fact that the fruits of his research have improved millions of lives.

Varun is also a promoter and an advocate of data science. In conjunction with colleagues, he runs ML India to build India's data science ecosystem, and has also organized world's first data science camp for schoolchildren. Varun's work has been covered in *The Economist*, *MIT Technology Review*, *The Wall Street Journal*, *HT Mint*, *The Economic Times*, and *IEEE Spectrum*. He has a passion for writing about science. His earliest works include an evaluation of Jagadish Chandra Bose's contribution to the invention of radio, which won him the Antique Wireless Association (AWA) award. He also writes poems and stories.